W9-BEO-857

# Theories of the World from Antiquity to the Copernican Revolution


by Michael J. Crowe

*University of Notre Dame*



DOVER PUBLICATIONS, INC.

*New York*

*To*
*Patricia Clare Crowe*
*and*
*Catherine Mary Crowe*



Published in Canada by General Publishing Company, Ltd., 30 Lesmill Road, Don Mills, Toronto, Ontario.
Published in the United Kingdom by Constable and Company, Ltd.

*Theories of the World from Antiquity to the Copernican Revolution* is a new work, first published by Dover Publications, Inc., in 1990.

Manufactured in the United States of America
Dover Publications, Inc., 31 East 2nd Street, Mineola, N.Y. 11501

*Library of Congress Cataloging-in-Publication Data*

Crowe, Michael J.
Theories of the world from antiquity to the Copernican revolution / by Michael J. Crowe.
p. cm.
Includes bibliographical references.
ISBN 0-486-26173-5
1. Astronomy—History. I. Title.
QB15.C77 1990
520′.9—dc20 89-25790
CIP

# Preface

*What This Book Is, and What It Is Not*

One way to describe this book and its limitations is to offer, as was common among eighteenth-century authors, an extended title: *A Selective History, Employing Elementary Geometrical Methods and Passages from the Writings of Ptolemy and Copernicus, of Theories of the Planetary System from Antiquity to 1615.* A briefer if more ambiguous title would be: *Given the Evidence in 1615, Which Theory of the Planetary System (the Ptolemaic, Copernican, Tychonic, etc.) Was Most Deserving of Support at That Time?* As will be suggested subsequently, answers to this question have significant historical and philosophical implications.

A second way of characterizing the contents of this volume is to provide an account of its origin. Around 1976, after teaching these materials for fifteen years at the University of Notre Dame, I began to compile portions of them into class handouts. Within a decade they had grown into a book-length manuscript, prepared without any idea that it might eventually be published. A summer's support from a National Endowment for the Humanities grant to Notre Dame for course development provided the opportunity to make a thorough revision of the materials and to put them into a word processor. The need for such materials arose from the fact that, although a number of excellent books cover this topic, none quite suited the needs of my students, who, although not yet ready for the technicalities treated in, say, Dreyer's classic *History of Astronomy from Thales to Kepler*, were nonetheless willing to explore these materials in a way that does not neglect their mathematical character. Fortunately, the mathematical techniques essential for a sound understanding of the great astronomical debate surveyed in this book are within the reach of anyone with good high-school training in mathematics and a willingness to use it. It also seemed advantageous to include readings from Ptolemy and Copernicus, who were the central figures in this debate. And it has proved to be an engaging approach to organize the materials so as to prepare readers to debate whether, given the evidence in 1615, they would have opted for a geocentric or heliocentric universe. Every year since

1976, these materials have been tested in the classroom and subsequently revised and updated. Not long ago, a colleague, Professor André Goddu, who has also used these materials in his classes, suggested that they be offered to a publisher. The ideal choice seemed Dover, which press has for decades provided readers interested in the history of scientific thought with books distinguished both by high-quality contents and low price. Dover's decision to publish it provided the opportunity for a final careful revision, designed to direct it to a larger public, while preserving those characteristics that have made it useful to the students the author has been privileged to teach.

This narrative of the origin of the book suggests to the reader that this volume is not a contribution to specialized research on the history of pre-1615 astronomy, for which many scholars possess superior credentials. It has rather been prepared to increase the accessibility of the writings of these scholars and of the great scientists whose writings they have so illuminated. The history provided is also selective, being organized chiefly around the aforementioned question.

*Three Points of View*

This book can be read more or less simultaneously from three different points of view: the scientific, the historical, and the philosophical.

If read for its **scientific** content, this book should provide a sound introductory knowledge of the **astronomy of our solar system**. The presentations of these ideas have been designed so as to draw, directly or indirectly, on the writings of a number of the most brilliant scientists of the past. Many of these ideas, being difficult, will challenge the reader's abilities and ingenuity. Nonetheless, being fundamental for an understanding of the world, they are essential to a liberal education. It is an irony of current educational practice that whereas everyone believes the earth orbits the sun, few persons can cite the evidences that led to this conviction. Numerous presentations of the astronomy of the solar system are available; what makes the following presentation different from most of these are two features. On the one hand, the materials are presented historically (more on this shortly). On the other hand, the materials are set out so as to give primacy to an understanding of the scientific ideas involved. Each idea is carefully explained in such a way that only a minimum of background information is necessary for its comprehension. The use of a mathematical approach in presenting these ideas derives from the conviction that the most clear, direct, and effective way to an understanding of not only the scientific, but also the historical and philosophical features of the Copernican revolution, is through mathematics. This is not to say that mathematics alone is sufficient; it is only to suggest that Copernicus was correct when in the preface to his great work, he stressed that in a fundamental way, mathematics is necessary for an appreciation of his arguments. One irony of the study of the Copernican revolution as herein presented is that it is only by studying it from a mathematical point of view that one can fully understand the crucial historical point that mathematics alone was insufficient for both its creation and its resolution.

Second, the materials can also be read for their **historical** content. The **Copernican revolution** ranks as one of the most important developments in the entire history of thought. Moreover, the story of how humans came to create and then abandon the Ptolemaic system is among the most interesting and dramatic narratives in all of history. The reader should be aware that in the following materials, this story is not presented in its full richness; in particular, many individuals and events, important in their own right but not essential to the mainline of this drama, have not been included. To cite two examples, the impressive astronomical methods developed in antiquity by the Babylonians have been passed over with only the briefest mention, and the significant advances in astronomy during the medieval period have also received less attention than would be expected in a study claiming comprehensiveness. The main justification for such omissions is that the materials have been selected so as to provide a knowledge of developments involved in a single grand-scale episode in the history of astronomy: the Copernican revolution. The goal behind this method of proceeding is to involve the reader as fully and directly as possible in this single episode, to provide an understanding of the ideas and problems of the individuals who played the most crucial roles in that episode.

One advantage of a historical approach is suggested by a passage from the writings of the German historian and philosopher Wilhelm Dilthey (1833–1911). In discussing the uses of history, Dilthey presented an idea that, although phrased in terms of religious history, applies with equal force to the history of scientific thought. He suggested that acquisition of historical experience can offset an unfortunate feature of our lives—that as we grow older, the range of our experiences becomes ever more limited. Nonetheless, Dilthey stated,

> . . . when I run through Luther's letters and writings, the reports of his contemporaries, the records of the religious conferences and councils as well as of his official communications, I live through a religious process of such eruptive power, of such energy, in which it is a matter of life or death, that it lies beyond any possibility of personal experience for a man of our day. But I can relive it.[1]

In short, these materials are designed to offer the reader an opportunity to relive to some extent one of the most dramatic developments in the history of thought. The chance of achieving this result will be enhanced by a commitment to understanding the ideas and arguments formulated by the pioneering astronomers and discussed in this book. The decision to adopt this approach has suggested some practices that may seem strange to twentieth-century readers. For example, what are now called the "Keplerian Laws of Motion" are herein referred to as the "Keplerian Conjectures," which more adequately characterizes how most of Kepler's contemporaries, at least those who took any notice of them, conceived them. To put this approach in different terms, we shall be examining both one **product** of science (the theory that the earth orbits the sun), and also the **process** by which that theory was attained. This approach, in which both the product and process of science are examined, can provide a fresh and more humanized view of science. We shall encounter not only scientific creations, but also scientific creators, not only settled conclusions, but also intense controversies. The restriction of the scientific

[1]Wilhelm Dilthey, *Gesammelte Schriften*, vol. 7, 2nd ed. (Leipzig, 1942), pp. 215–16.

ideas in this book to those available in 1615 has not precluded presentation of historical information that has become available since that year. An example of this is the Appendix to the book, which consists of a presentation of very recent theories of the astronomical functions of such megalithic sites as Stonehenge, even though no historian alive in 1615 possessed a knowledge of these theories. Readers uninterested in these theories can omit this presentation without losing anything other than some illustrations of the information contained in Chapter One.

Third, this book presents various **philosophical** ideas and issues. The Copernican claim that the sun, not the earth, is the center of our system presented many challenges to traditional philosophical and theological beliefs. A number of these challenges are discussed in the book and illustrated in its Epilogue, which consists of a collection of quotations from various authors who pondered these issues with special intensity. Moreover, the Copernican revolution raises an array of methodological issues, such as the nature of scientific discovery and verification, the status of theoretical entities, and the role of empirical evidence. The two methodological questions that emerge most strikingly from these materials are (1) what is the relation of scientific theories to empirical information, and (2) what approach should be taken concerning the nature and testing of scientific theories? Francis Bacon is among those who have urged in regard to the first issue that scientific theories arise from empirical information; in fact, his writings seem at times to convey the claim that the necessary and sufficient basis for the creation of new scientific theories is detailed knowledge of the phenomena. Other philosophers have challenged this view, which is frequently termed "inductivism," urging that empirical factors are usually insufficient for either the creation or adequate testing of scientific theories. A crucial question involved in the second issue is whether theories should be viewed as primarily instruments, as useful fictions devised to account for the world and to predict future events, or whether they should be construed in a realist sense, as hopefully true portrayals of nature that can be expected not only to account for and to predict natural phenomena, but also to explain them. These issues are fundamental for far more than astronomy; they are, in fact, relevant to almost every area of learning.

*Acknowledgments*

I am, first of all, deeply indebted to the many students at the University of Notre Dame who have not only inspired the preparation of these materials, but also by their questions and comments have increased their clarity and accessibility. Special thanks are due to one of the most recent of these, Jamie Brummer, who located many of the literary statements concerning the Copernican revolution that appear in the Epilogue. Among colleagues at Notre Dame, much appreciation is due to Professors Frederick J. Crosson, André Goddu, Phillip R. Sloan, and Patrick Wilson. To the inspired teaching of the first, I owe my initial contact with the Copernican revolution. To the generous good will of the second, I am indebted not only for many revisions, but also for his encouragement (which was crucial) for the idea of publishing these materials. My indebtedness to the third includes his support for essentially all aspects of this project. Professor Wilson's careful reading of the entire manuscript and insightful suggestions concerning its improvement leave me deep in his debt. Very helpful comments were also received from colleagues at other institutions, including Professors William Ashworth, James Evans, Harvey Flaumenhaft, Stephen T. Gottesman, and Robert A. Hatch. Warm thanks, also, to John W. Grafton of Dover Publications for his encouragement of this undertaking.

I am indebted to the following persons and publishing companies for permission to quote or reprint materials from the sources indicated: Barry M. Casper and Richard J. Noer for use of a diagram from their *Revolutions in Physics* (New York: W. W. Norton, 1972); British Heritage for a photograph and diagram of Stonehenge; Cambridge University Press for translations in M. J. Crowe, *The Extraterrestrial Life Debate 1750–1900: The Idea of a Plurality of Worlds from Kant to Lowell* (Cambridge, England, 1986), for a translation in Steven J. Dick, *Plurality of Worlds: The Origins of the Extraterrestrial Life Debate from Democritus to Kant* (Cambridge, 1982), and for a passage from G. W. Leibniz, *New Essays on Human Understanding*, tr. Peter

Remnant and Jonathan Bennett (Cambridge, 1982); E. C. Krupp and the Griffith Observatory for a map from E. C. Krupp, *In Search of Ancient Astronomies* (Garden City, New York: Doubleday & Company, 1978); G. J. Toomer and Gerald Duckworth and Company Ltd. for a selection from *Ptolemy's Almagest,* tr. G. J. Toomer (New York: Springer-Verlag, 1984); Johnson Reprint Company for passages from *Kepler's Conversation with Galileo's Sidereal Messenger,* tr. Edward Rosen (New York, 1965); Mouton De Gruyter & Co. for a passage from Giordano Bruno, *The Ash Wednesday Supper,* tr. Stanley L. Jaki (The Hague, 1975); Polish Academy of Sciences for the selection from Nicholas Copernicus, *On the Revolutions,* ed. Jerzy Dobrzycki and tr. Edward Rosen (Baltimore: Johns Hopkins University Press, 1978); Princeton University Press and Routledge & Kegan Paul Ltd. for a translation given in S. Sambursky, *The Physical World of Late Antiquity* (London, 1962); Robert E. Krieger Publishing Company, Inc., for translations in Edward Rosen, *Copernicus and the Scientific Revolution* (Malabar, Florida, 1984); Routledge & Kegan Paul Ltd. for a passage from G. W. Leibniz, *Theodicy: Essays upon the Goodness of God, the Freedom of Man, and the Origin of Evil,* tr. E. M. Huggard (New Haven, 1952); Stillman Drake and Doubleday & Company Inc. for translations from Stillman Drake, *Discoveries and Opinions of Galileo* (Garden City, New York, 1957); University of Chicago Press for translations given in Pierre Duhem, *To Save the Phenomena,* tr. Edmund Doland and Chaninah Maschler (Chicago, 1969), for passages from Moses Maimonides, *The Guide of the Perplexed,* ed. and tr. Shlomo Pines (Chicago, 1963), and for passages from Michael Polanyi, *Personal Knowledge* (New York: Harper & Row, 1964).

Michael J. Crowe
University of Notre Dame

# *Table of Contents*


| | | |
|---|---|---|
| Preface | | |
| Chapter One | The Celestial Motions | 1 |
| Chapter Two | Greek Astronomy before Ptolemy | 23 |
| Chapter Three | Some Mathematical Techniques of Ancient Astronomy | 32 |
| Chapter Four | The Ptolemaic System | 45 |
| Chapter Five | Philosophical Interlude: The "Save the Phenomena" Position | 69 |
| Chapter Six | The Copernican System | 85 |
| Chapter Seven | The Tychonic System | 137 |
| Chapter Eight | Johannes Kepler | 147 |
| Chapter Nine | Galileo Galilei | 157 |
| Epilogue | Some Quotations Concerning Astronomy in the Copernican Revolution Period | 174 |
| Appendix | Archaeoastronomy | 198 |
| Bibliography | | 221 |
| Index | | 227 |

# Chapter One

## *The Celestial Motions*

### *The Motion of the Stars*

This chapter doubly challenges the imagination. First, it draws on the geometrical or spatial imagination needed to conceive of the motions of the stars, sun, moon, and planets. Second, this chapter, nearly all of which could have been written by an ancient Greek astronomer, invokes the historical imagination by presenting these motions in the way that the Greeks envisioned them, that is, from a geocentric (earth-centered) perspective, which is, of course, how we see them. This approach will not only assist in understanding some of the ancient astronomies, but also facilitate a comprehension of these motions as conceptualized from the modern heliocentric (sun-centered) point of view.

Persons watching the **stars** over a number of nights see that nearly all of them appear to move in a counterclockwise direction along circles varying in size.

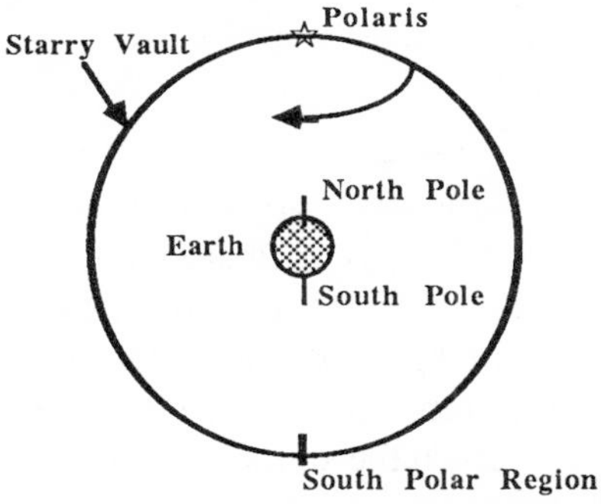


The sole stationary stars are **Polaris** and the southern polar stars. The motions of the stars are identical to what they would be if they were all located on a huge sphere, the **starry vault**, rotating once approximately every twenty-four hours, and having as its center the earth, which is assumed to be motionless. The sense of rotation of the starry vault is such that a star on its right side is moving out of the page. Typically, a given star will appear to rise on the eastern horizon and set on the western horizon.

How do persons living on the earth's equator see the stars

move? Polaris and the southern polar stars appear fixed in position. The remaining stars rise perpendicularly to the eastern horizon and set perpendicularly to the western horizon. These motions are represented in the next diagram in which the ellipse represents the horizon plane.

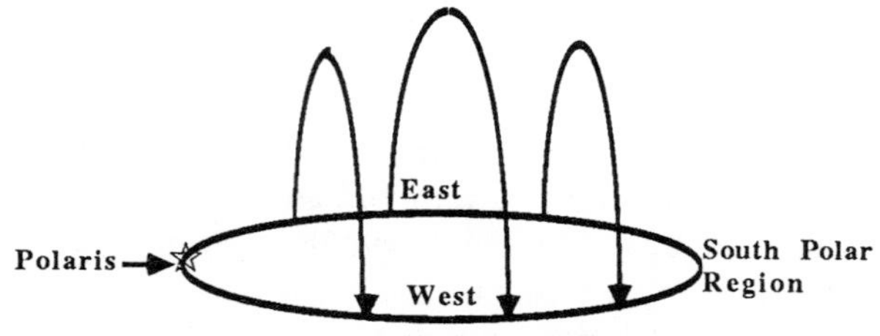


Horizon Plane at Equator

At the north pole, Polaris is seen fixed in position at the **zenith** (the point in the heavens directly above a person's location). The other stars appear to move in circles parallel to the horizon plane and centered on Polaris.

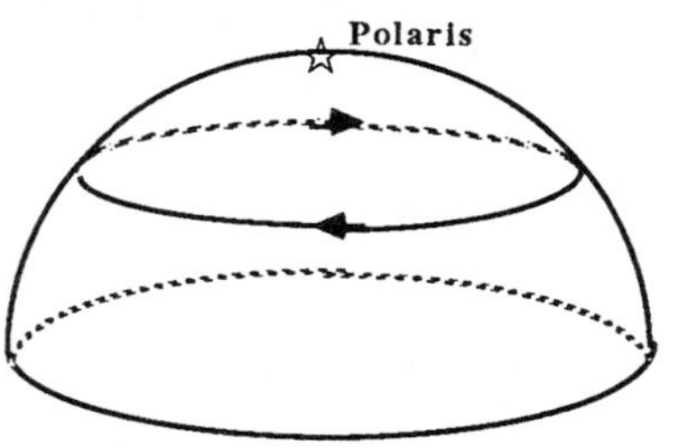


Horizon Plane at North Pole

Persons living in Chicago or Boston are located at 42° north terrestrial latitude, i.e., 42° up from the equator. As the next diagram indicates, for such persons, Polaris appears fixed in position, whereas the stars near it move in circles that are always visible. Stars farther down the starry vault move in circles that are cut by the horizon plane.

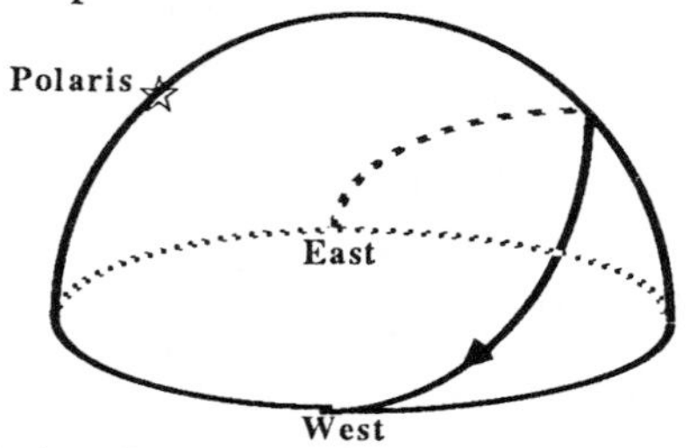


Horizon Plane at 42 Degrees North Latitude

*Problems*

**Problem 1**: Polaris is the last star in the handle of the little dipper. Draw the motion of the stars in the Little Dipper as seen from 42° north latitude over a period of three hours.

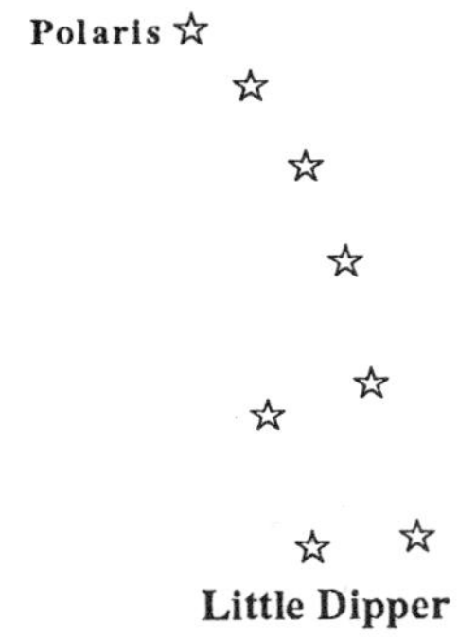


**Little Dipper**

**Problem 2**: Suppose that there exist only two bodies in the universe: one is identical to the starry vault and the other is a very small spherical planet located at the center of the starry vault. Let us assume that the starry vault rotates once every 24 hours, whereas the planet remains fixed in position, i.e., it neither rotates nor revolves. What motion would an inhabitant of the starry vault, convinced that the starry vault is at rest, attribute to the planet at the center? Represent this by means of a diagram. Are there any conclusive arguments that an inhabitant of the planet could formulate to prove that the starry vault is rotating? Or are there arguments that the inhabitant of the starry vault could use to show that the planet is rotating?

## *The Motion of the Sun*

First, some definitions:

**Celestial Equator:** The celestial equator is the line on the starry vault that lies directly above the earth's equator. Some point on the celestial equator is always at the zenith of a person living on the earth's equator.

**Ecliptic:** The ecliptic is a line on the starry vault on which the sun always appears to be located. It is the apparent **yearly** path of the sun or the projection of the path of the sun on the starry vault during one year. The ecliptic is inclined at 23 1/2 degrees to the celestial equator.

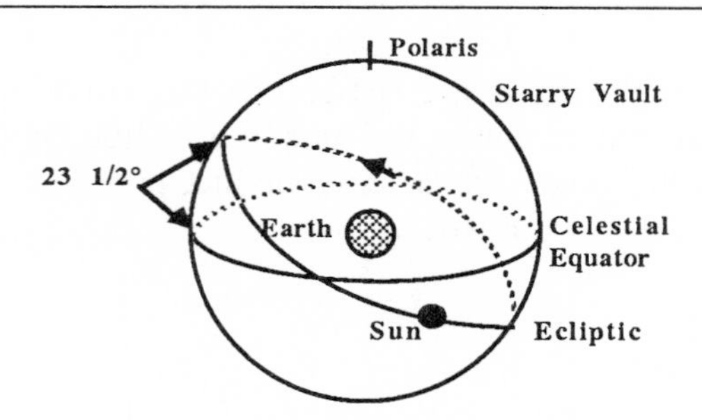


The sun completes one circuit of the ecliptic every **365.24220 days**. From this it is evident that the sun moves approximately one degree each day on the ecliptic. Note that whereas the stars move from **east to west**, the motion of the sun on the ecliptic is from **west to east**. Let us now combine the motion of the starry vault with that of the sun. It is important to remember that the ecliptic is simply a line among the stars; it is like a seam on a basketball. If a basketball is rotated, its seam rotates with it. Correspondingly, the ecliptic rotates with the starry vault. Consequently, each day the sun makes one revolution around the earth along with the starry vault; however, the sun also moves about one degree per day along the ecliptic, moving in the opposite direction. As an aid to visualizing this, imagine an ant walking slowly down the side of a rapidly rotating basketball. The path of the ant from the point of view of the basketball is a straight line, but if seen from a fixed observer at a distance, the ant will appear to be moving along a helix. The next diagram shows the motion of the sun for one day.

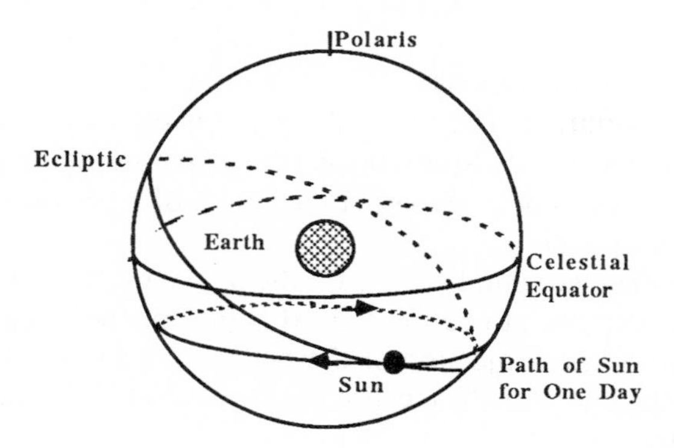

Additional definitions are now needed.

**Vernal Equinox:** The point on the starry vault where the ecliptic crosses the celestial equator with the sun moving toward the northern half of the heavens. The sun is at the vernal equinox around March 21. When the sun is at an equinoctial point (vernal or autumnal equinox), people on earth in most cases experience days and nights of equal length.

**Summer Solstice:** The most northerly point on the ecliptic. The sun is at the summer solstice around June 22.

**Autumnal Equinox:** The point on the starry vault where the ecliptic crosses the celestial equator with the sun moving toward the southerly half of the heavens. The sun is at the autumnal equinox around September 23.

**Winter Solstice:** The most southerly point on the ecliptic. The sun is at the winter solstice around December 22.

These definitions can be presented diagrammatically.

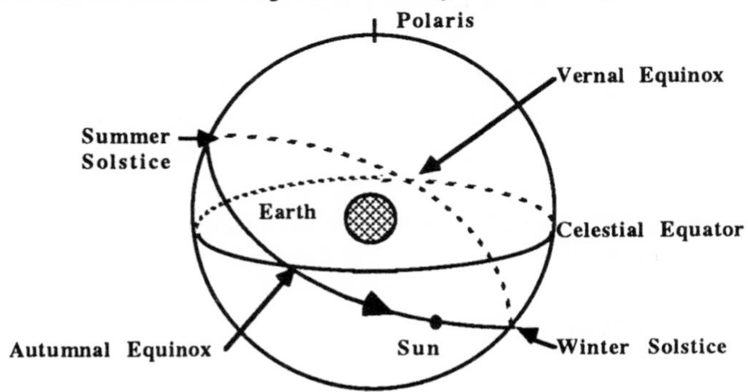


*The Seasons*

The discussion of the sun's motion presented up to this point can be used to explain the seasons. This will also provide practice in applying these ideas. Many inhabitants of the northern hemisphere believe that summer is hot because the sun is closer at that time than in winter. In fact, the sun is farther from the earth in summer than in winter. The chief reason why summer is warm and winter cold is the difference between the angles at which the sun's rays reach us during those seasons. Let us imagine the sun on the celestial equator and determine the angle at which the sun's rays strike such cities as Chicago or Boston, which are located at

42° N. terrestrial latitude. As the diagram shows, when the sun is at an equinoctial point, it is directly above the equator. Hence its rays strike a point at 42° N. latitude at an angle of 42°. Thus at the vernal and autumnal equinoxes, solar rays arrive at Chicago at an angle of 42° off the vertical.

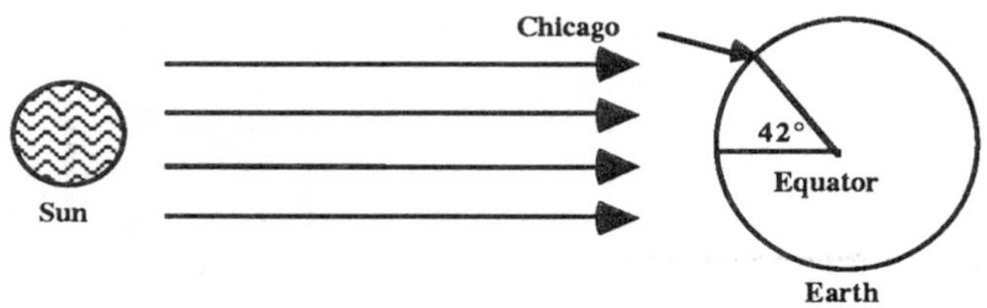


When the sun is located at the summer solstice, it is 23 1/2° up from the equator, which causes its rays to strike Chicago at 42° – 23 1/2° = 18 1/2° from the vertical. Consequently, in summer the sun's rays come closer to the perpendicular than in fall or spring. Many people in the northern United States believe they have seen the sun directly overhead; in fact, the sun never gets closer to being directly overhead than at the time of the summer solstice and even then it is 18 1/2° from the zenith at 42° N. latitude. The next diagram shows the relative positions of the sun and earth at the time of the summer solstice. Note the difference between this and the previous diagram.

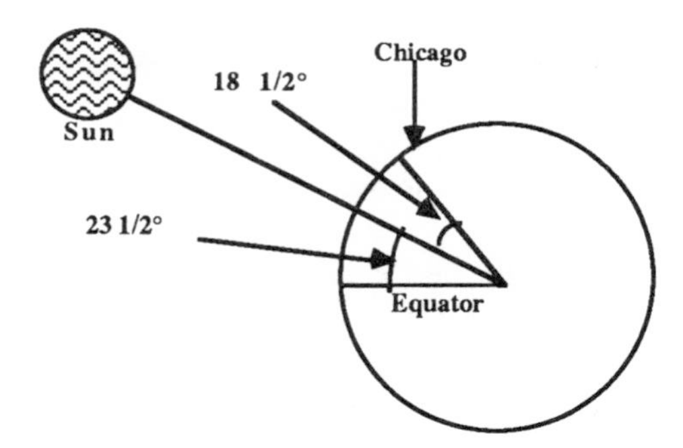


When the sun is at the winter solstice, it is located 23 1/2° below the equator. Consequently, its rays strike Chicago at 42° + 23 1/2° = 65 1/2° from the vertical. This is to say that when the sun is at the winter solstice, it is at most 24 1/2° (90° – 65 1/2°) above the horizon.

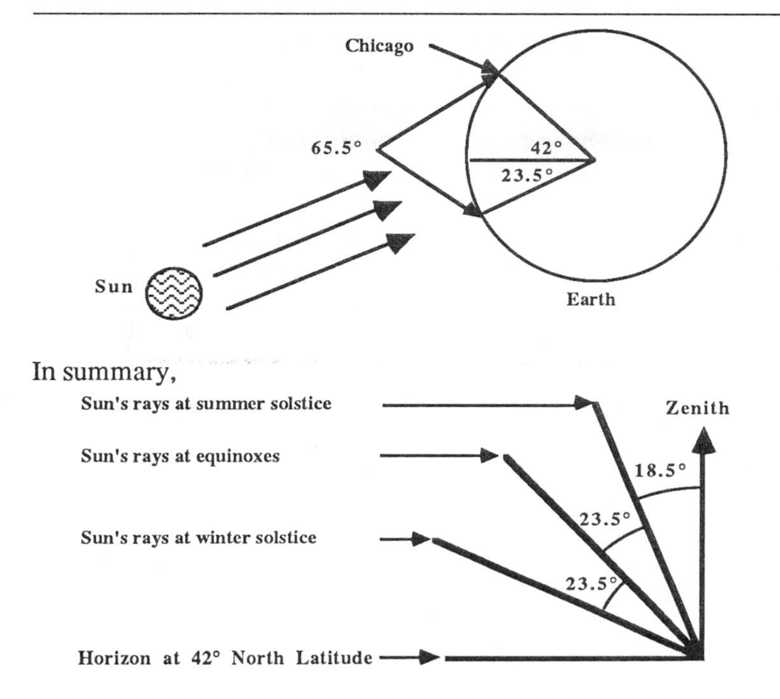


In summary,

From this one sees that the greatest heating of a region occurs when the sun's rays strike it most directly, whereas when the sun's rays hit it more obliquely less heating occurs. This is the chief factor influencing our seasons.

The following problems illustrate the materials presented up to this point.

**Problem 3**: Determine the length of the night that began on September 23, 1845, in Clausville, which is located at 134° east terrestrial longitude and 90° north terrestrial latitude.

**Problem 4**: Given below is a plot of temperature versus month for the northern continental United States.

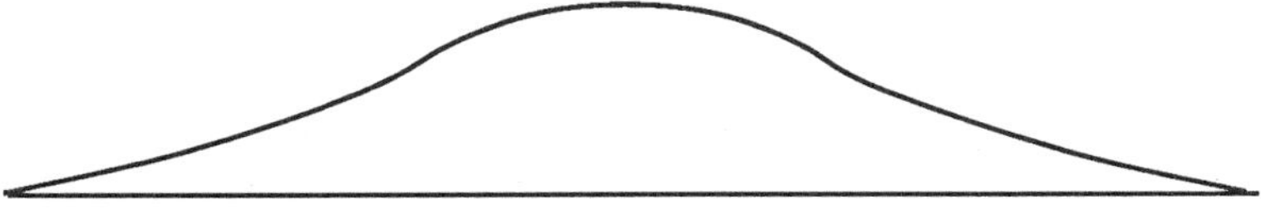


Draw a comparable plot of temperature versus month for (a) persons living in Sydney, Australia (about 35° south latitude), and for (b) persons residing in Quito, Ecuador, which is located on the equator.

**Problem 5**: The tropical or seasonal year, i.e., the average period between the beginning of spring in one year and that in the next, is 365.242200 days. This is the basic factor that must be kept in mind in devising a calendar. For each of the calendars proposed below, calculate the amount of error that will be introduced by that calendar in a period of two thousand years.

Calendar I: The year is 365 days long.

Calendar II: The year is 365 days long, unless the year number is divisible by 4, in which case the year is 366 days long.

Calendar III: The year is 365 days long, unless the year number is divisible by 4, in which case the year is 366 days long. If, however, the year number is divisible by 100, but not by 400, the year remains 365 days long.

**Problem 6**: Suppose only two bodies, **A** and **B,** exist in the universe. Suppose that the inhabitants of **B** see **A** move around **B** in a circle in one year.

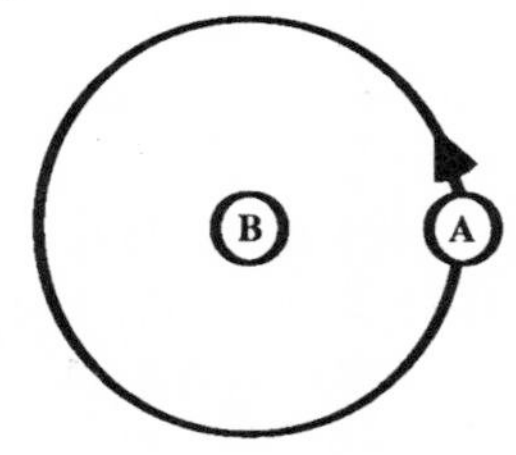


Let us suppose that inhabitants of **A** believe that **A** is motionless. Specify the period and the shape of the orbit that they will attribute to **B**. Let us assume that the body **A** always retains the same orientation; that is, were there actually a letter **"A"** on top of it, that letter would continue to remain right side up. Put another way, assume that body **A** does not rotate. Are there any arguments that inhabitants of **B** could use to prove to inhabitants of **A** that it is their body that is actually revolving?

*Precession of the Equinoxes*

The motions of the stars and sun are more complicated than presented so far. Another important aspect of the heavens, the **precession of the equinoxes**, was known to the Greeks who found that the equinoctial points, the points where the ecliptic crosses the celestial equator, change slightly; specifically, each equinoctial point makes a full circuit around the ecliptic in **26,000 years** (the Greeks thought 36,000 years). To visualize this, imagine two lines running through the earth, one perpendicular to the plane of the celestial equator, the other perpendicular to the plane of the ecliptic. These lines extend respectively to the pole of the celestial equator (currently Polaris) and to the pole of the ecliptic. As the diagram shows, the polar line for the celestial equator turns around the ecliptic's polar line in 26,000 years. Thus the celestial equator slowly changes position over this period.

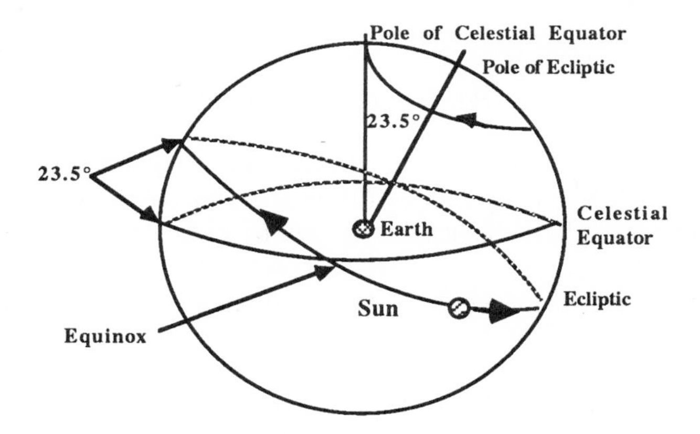


A major result of precession is that Polaris is ceasing to be our pole star. In 13,000 years, a star 47° (twice 23 1/2°) from Polaris will be our north star.

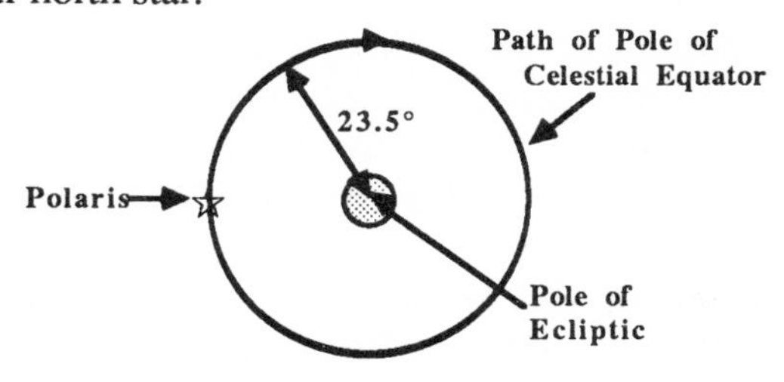


This phenomenon can also be visualized by imagining that the earth rotates around an axis that turns through a circle of radius 23 1/2° in a period of 26,000 years.

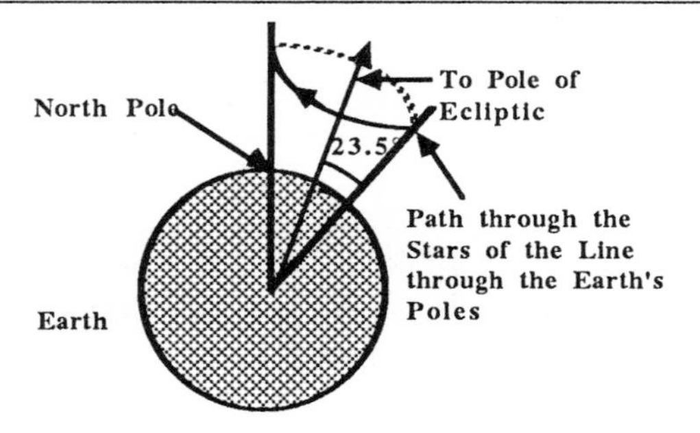


The detection of the precession of the equinoxes was one of the most important observational results achieved by ancient astronomers.

*The Motions of the Moon*

The analysis and prediction of the moon's motions presented a major challenge to ancient civilizations. One complexity in this regard derives from the fact that the nearness of the moon makes it possible to observe her motions with far greater precision than for most other celestial bodies. This challenge was also felt very strongly because of the importance of the moon's motions. An understanding of those motions was crucial not only for knowing when the moon would provide nocturnal illumination, but also for predicting eclipses, which were viewed as spectacular and very significant phenomena, and for constructing calendars, which in ancient times were frequently lunar. In what follows, the main motions of the moon will be examined in detail, not only because they are important in their own right, but also because an understanding of these motions is essential in any assessment of the claims made in recent decades that Stonehenge and other megalithic sites were designed as astronomical observatories.

First, some definitions:

**Sidereal Period of the Moon:** This is the time it takes the moon to orbit the earth and to realign itself with a star. It is equal to **27.32166 days.**

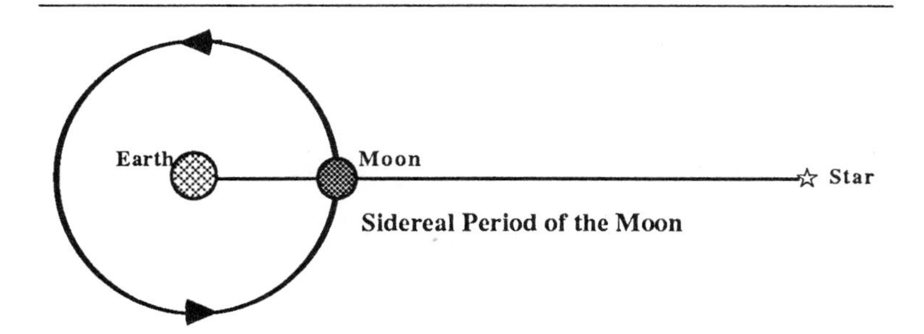


It is very important to distinguish the sidereal period of the moon from its synodic period.

**Synodic Period of the Moon:** The moon's synodic period is the time it takes the moon to complete one orbit and to continue on to the point that it is realigned with the sun. It equals **29.53059 days**, i.e., the time between successive new or full moons.

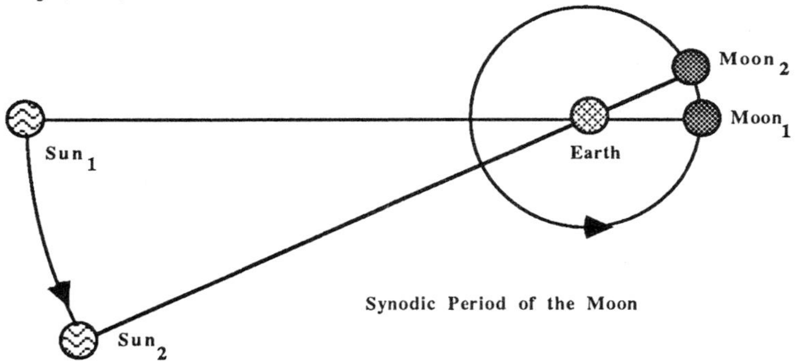


**Zodiac:** This is the region reaching 8° on either side of the ecliptic. The motions of the sun, moon, and the classical planets are confined to the zodiac. It is divided into twelve constellations or houses.

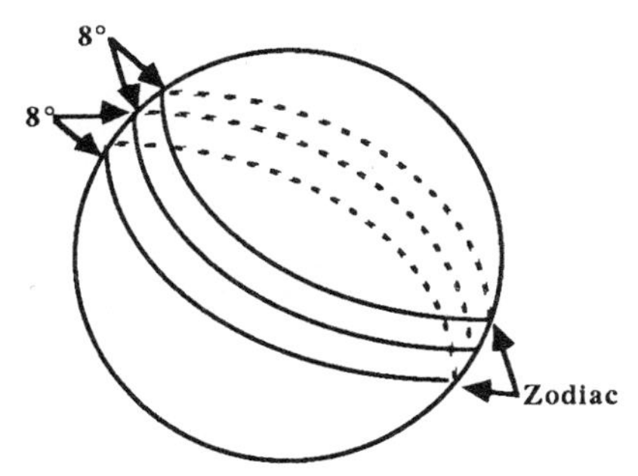

Were the zodiac to be unwrapped like a belt and laid on a page, the moon's orbit would appear as shown in the next diagram, from which it is evident that the moon's orbit is inclined at 5° to the ecliptic.

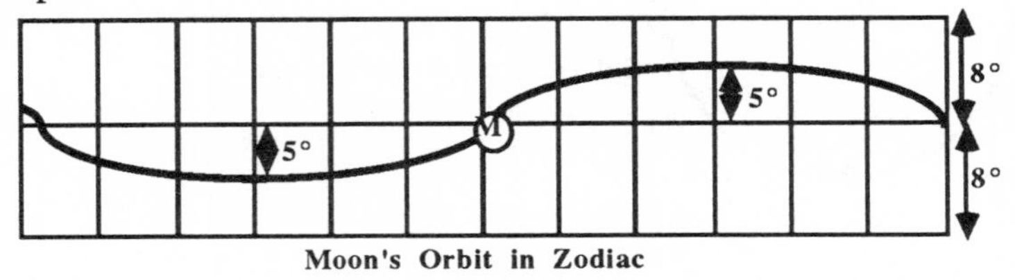

Moon's Orbit in Zodiac

*The Phases of the Moon*

As is well known, the moon shows various phases.

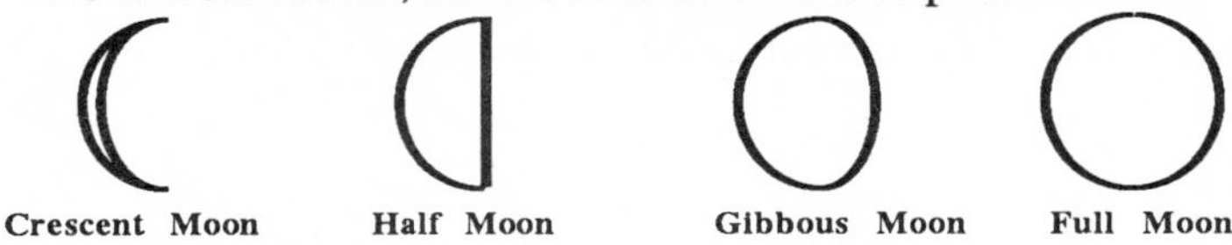


The Greeks realized that the moon's phases do not result from the earth blocking the sun's light from the moon. Rather, as the next diagram shows, the phases are a result of how the moon is seen from earth. We always see half the moon (the side nearer to us) and half the moon is always illuminated by the sun (the side toward the sun). Depending on the time in the lunar month, we see all the moon's illuminated half or only a portion of it.

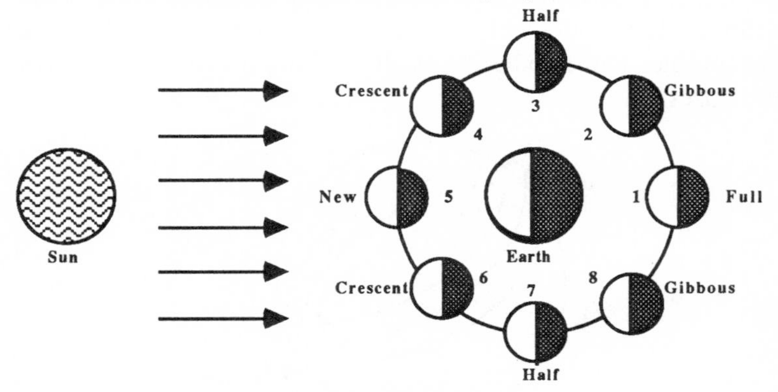

The Phases of the Moon

When the moon is in positions 4 or 6, it is seen as a crescent; when in positions 3 or 7, it appears as a half moon. At positions 2 or 8, it is in its gibbous phase. When at 1, it is seen full. When at position 5, the moon cannot be seen, both because of the sun's glare and because the moon's dark side faces us.

*Eclipses*

Eclipses are among the most dramatic celestial events. Two types of eclipses concern us: solar and lunar.

**Solar Eclipse:** Occurs when the moon comes between the earth and sun, the moon blocking the sun's rays from reaching the earth.

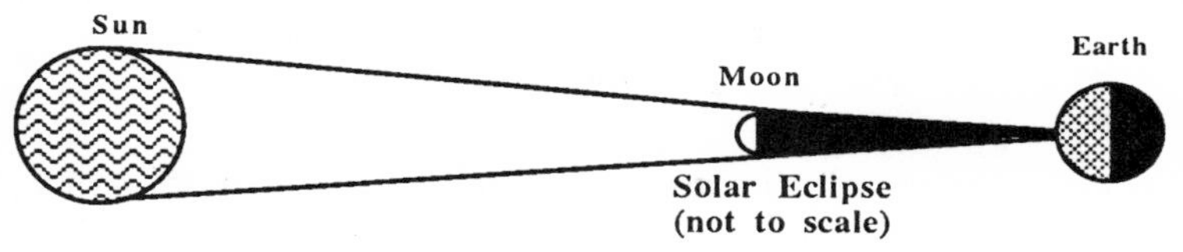

Solar Eclipse
(not to scale)

**Lunar Eclipse:** Occurs when the earth comes between the sun and moon, blocking the sun's rays from reaching the moon.

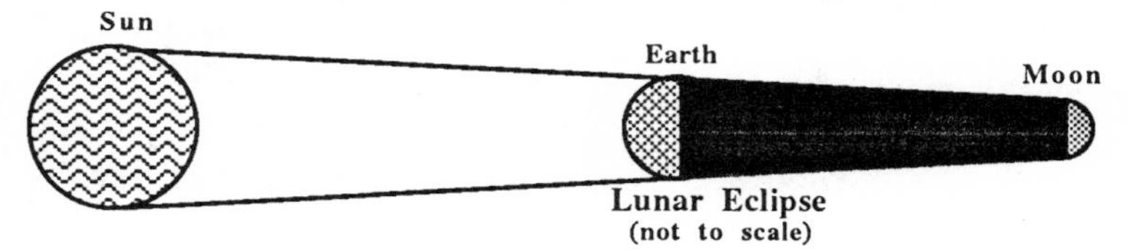

Lunar Eclipse
(not to scale)

Because of the geometry of the two types of eclipses, solar eclipses are seen over only a very limited region of the earth, whereas lunar eclipses (which entail an actual darkening of the moon) are seen identically from all portions of the earth from which the moon can be seen at the time of the lunar eclipse. Let us now investigate the conditions necessary for eclipses of each type to occur.

One might think that a solar eclipse would occur once for each synodic period (29.53059 days), i.e., every time the sun and moon reach the same portion of the zodiac. This does not, however, occur, the reason being that the moon's orbit is inclined at 5° 9' (or 5.14°) to the plane of the sun's orbit. Because of this, when the moon passes the sun, it is usually above or below the sun, and although it blocks out the rays of the sun, the resulting shadow cone misses the earth. The point where the moon, moving upward, crosses the ecliptic is called the "**ascending node**"; the point where the moon, moving downward, crosses the ecliptic is known as the "**descending node**."

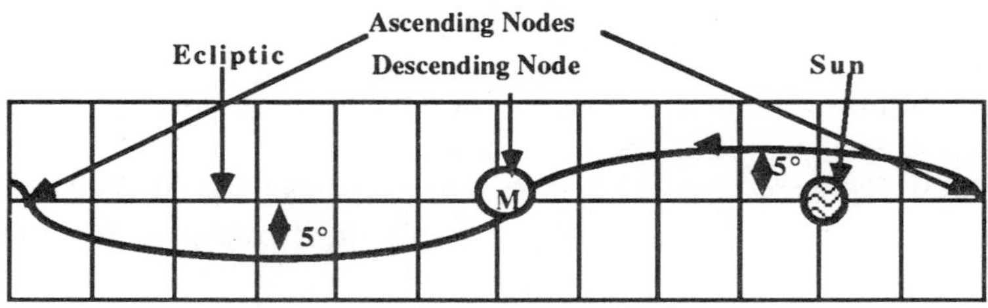


**Moon's Orbit in the Zodiac, Showing Positions of the Nodes and of the Sun**

As this diagram suggests, for a solar eclipse to occur, the moon must be on or very near the ecliptic and in the same region of the zodiac as the sun. For a lunar eclipse to occur, the moon must be on or near the ecliptic and 180° distant on the zodiac from the sun. An understanding of this relationship makes clear why if a solar eclipse occurs, it is probable that a lunar eclipse will follow it, occurring about 14 days later.

It is important to note the time between the moon's passing through successive descending nodes is slightly shorter than its sidereal period. This shorter period, known as the moon's **draconitic period,** is illustrated in the diagrams by the fact that the moon makes one sweep up and down—and a fraction more—as it moves through the zodiac. The moon's draconitic period is **27.2122 days**, whereas the sidereal period of the moon is 27.32166 days.

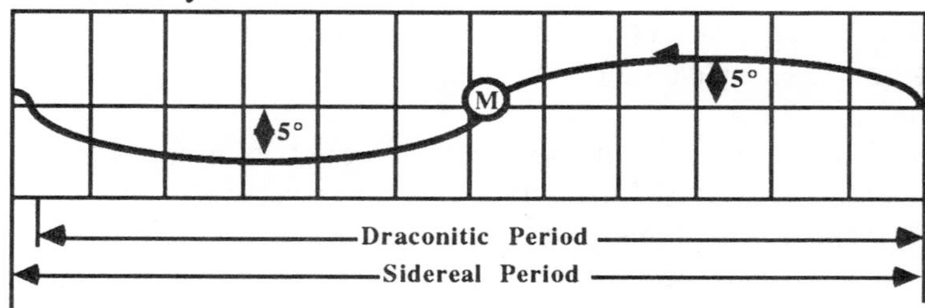

*Problems*

Hint: Finding solutions to most of these problems will be facilitated by beginning with a diagram that incorporates the information provided in the problem.

**Problem 7**: Suppose that on the night of August 13, 1301, a lunar eclipse occurred that was seen in London. What was the phase of the moon on the night of August 12, 1301? Explain your reasoning.

**Problem 8**: Suppose that on November 20, 1492, persons in Cairo, Egypt, saw a full moon. What percentage of the night was the moon visible that night? What percentage of the night was the moon visible November 11, 1492? Your first answer should be precise, but your second answer can be an approximation.

**Problem 9:** The attempt has sometimes been made to account for the darkening of the sky during Christ's crucifixion by attributing the darkening to a solar eclipse. According to the Synoptic Gospels, the crucifixion occurred on the afternoon after the Jewish Passover meal, which meal was set on the 15th day of the month Nisan. The Jewish calendar being a lunar calendar, the first day of each month coincided with a new moon. Use this information to comment on the acceptability of the proposed explanation.

**Problem 10**: Assume that both the moon and sun move in circular orbits around the earth. Assume also that the synodic period of the moon is exactly 30 days and that the sun goes around the earth in 360 days. On the basis of these simplified assumptions, calculate the sidereal period of the moon.

## *The Motions of the Planets*

One of the classic problems of astronomy has been to account for the motions of the planets. A planet looks rather like a star, except for two features: (1) whereas stars remain fixed in position and brightness relative to each other, planets move in relation to the stars and also change in brightness, and (2) planets tend to be brighter than most stars. There are five classical planets, all of which have been known from the beginning of recorded history; they are:

**Mercury**
**Venus**
**Mars**
**Jupiter**
**Saturn**

The planets, except in certain important cases noted in the next paragraph, have motions similar to those of the sun and moon; in particular, planets rise once every 24 hours with the stars on the eastern horizon and set in the west. Also, like the sun and moon, they gradually move through the zodiac in the direction from west to east. Whereas the sun is always on the ecliptic and the moon never more than 5° 9' from it, each planet deviates from the ecliptic by a particular amount, Mercury departing by as much as 7°.

Two peculiar features distinguish planetary motions from those of the sun and moon. First, whereas the sun and moon move through the zodiac independently of each other, two planets, Mercury and Venus, have **bounded elongation**; that is, Mercury and Venus are always found within a certain distance of the sun. In particular, Mercury is always found within **28°** on either side of the sun and Venus within **46°**. Because of this, Mercury and Venus are only seen in the hours after sunset or before sunrise; they are never visible during the hours of the night when the sun is farther than 46° below the eastern or western horizon. When Scripture or poets refer to the morning or evening star, Venus is usually meant, although Mercury (which is usually far less bright than Venus) may possibly be what the writer has in mind. Mars, Jupiter, and Saturn do not have bounded elongation. The second important feature of planetary motion is that all planets at certain times move in a retrogressive manner; that is, whereas they usually move, like the sun and moon, from west to east in the zodiac, at times the planets appear to back up. Such motion, which is called **retrograde motion**, is exhibited in the next diagram:

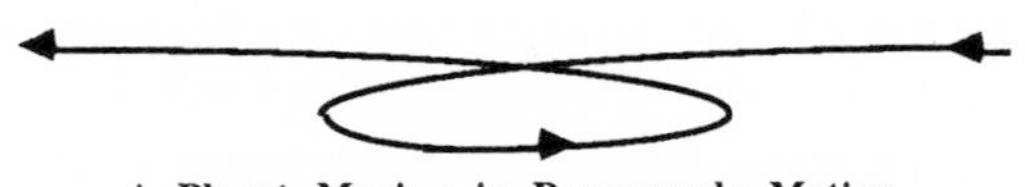

A Planet Moving in Regrograde Motion

In summary,

| | |
|---|---|
| **Have bounded elongation:** | **Mercury** |
| | **Venus** |
| **Have retrograde motions:** | **Mercury** |
| | **Venus** |
| | **Mars** |
| | **Jupiter** |
| | **Saturn** |

In discussing planetary motions, we shall need the following definitions.

**Zodiacal Period:** The zodiacal period of a planet is the average time that the planet takes to complete one circuit of the zodiac.

**Opposition:** A planet is said to be in opposition with the sun when the planet is on the opposite side of the earth from the sun, i.e., when it is 180° away from the sun's position on the zodiac. Because of their bounded elongations, Mercury and Venus are never in opposition with the sun.

A Planet in Opposition with the Sun (not to scale)

**Conjunction:** A planet is said to be in conjunction with another body (the sun, moon, or another planet) when it is in the same region of the zodiac as that other body. The planet may, however, be above or below the second object.

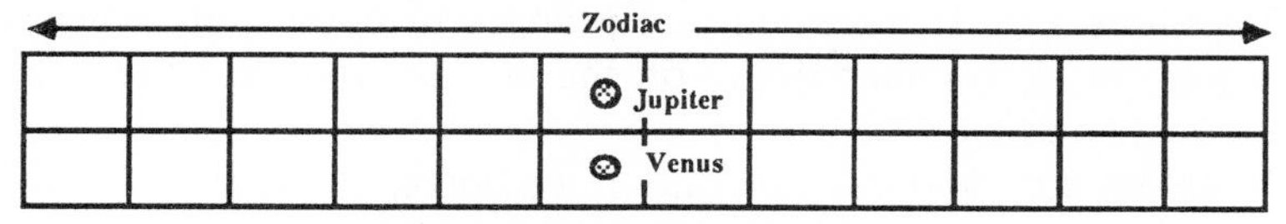


Jupiter and Venus in Conjunction

**Synodic Period of a Planet:** For Mars, Jupiter, and Saturn, the synodic period is defined as the average time between successive oppositions with the sun. For Mercury and Venus, the synodic period is the average time it takes them to pass through not one, but two conjunctions.

Given next is a table in which the zodiacal and synodic periods for each planet are specified.

| Planet | Zodiacal Period | Synodic Period |
|---|---|---|
| Mercury | 1 year | 115.08 days |
| Venus | 1 year | 583.92 days |
| Mars | 1.881 years | 779.94 days |
| Jupiter | 11.862 years | 398.88 days |
| Saturn | 29.458 years | 378.09 days |

Note the following points from the table:
(1) Saturn takes the longest time to travel through the zodiac. Correspondingly, it moves through the constellations, i.e., changes its position relative to the stars, only very slowly.
(2) Jupiter also has a very long zodiacal period.
(3) Mars moves faster than either Jupiter or Saturn.
(4) Both Mercury and Venus have the same zodiacal period, one year. This fits with the fact that they have bounded elongation. They must, on the average, travel through the zodiac with the same period as the sun.
(5) Were the sun included in this table, it would have a zodiacal period of one year. Because the synodic period of a celestial body is the time it takes for it to realign itself with the sun, it is not meaningful to specify a synodic period for the sun.
(6) No obvious pattern is apparent from a comparison of the synodic periods of the various planets.

*Problems*

**Problem 11**: If informed that Mars is in opposition, where would an astronomer living on the equator look for Mars at midnight on June 23?

**Problem 12**: Suppose that only three bodies A, B, and C exist in the universe (see diagram). Suppose that A and B move in circular orbits around C with periods $P_A$ and $P_B$ respectively and with $P_A > P_B$. Assume that A does not rotate, i.e., that it moves so that were the letter A actually affixed to it, A would always remain with its bar parallel to the bottom of the page.

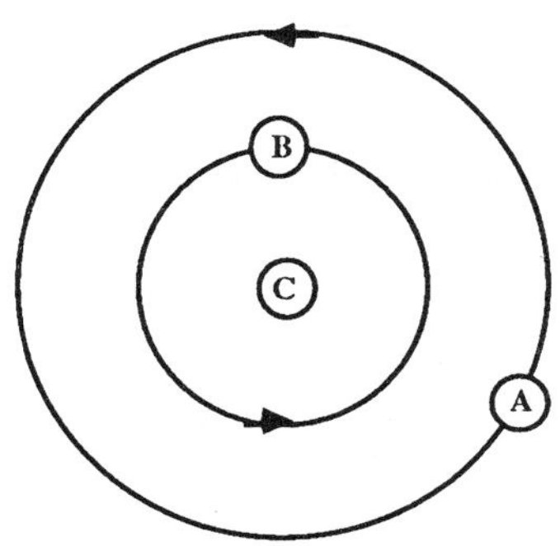


Draw a diagram composed of one or more circles showing how a person living on A would see the motion of C; what path and what period would it have? Second, describe how a person living on A would see the motion of B; i.e., again restricting yourself to one or more circles, construct a model of the motion of B in relation to A. What period and what path would B have? How would its motion be related to C's motion? Finally, state whether there are any reasons that an inhabitant of C could use to convince an inhabitant of A that it is A that is actually moving.

**Problem 13**: (Optional) Suppose A, B, and C are the only three bodies in the universe and suppose that they move as in problem 12. Using one or more circles, describe how an inhabitant of B would see the motions of A and C; determine the periods and paths that the inhabitant of B would attribute to A and C. Are there any convincing reasons that the inhabitant of B could use to convince the inhabitants of A or C that they are actually moving?

## *Conclusion*

These are the main motions of the stars, sun, moon, and planets. As will be evident shortly, ancient astronomers showed remarkable skill in devising theories and applying mathematical devices to account for these motions.

*Appendix: Discussion of Some Features of Circular Motion*

Suppose that three bodies move as indicated in the next diagram. Body C is fixed in the center, whereas B moves around C with period $P_B$. Similarly, A moves around C, but in a larger circle and with a longer period, $P_A$. Suppose that the bodies B and A always retain the same orientation in relation to the page, e.g., someone on the left side of B remains on the left side of B as B moves through its circle.

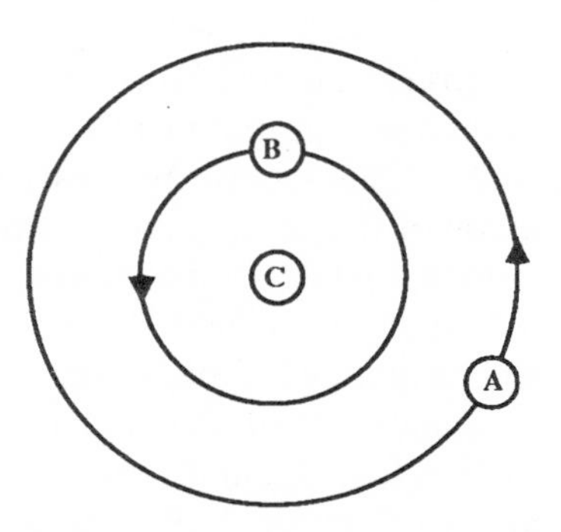


Let us first investigate how someone on B will see the motions of C and A. It is clear that a person on B will see C as remaining always at the same distance (CB) from B. Moreover, C will have a period equal to $P_B$. The direction of C's motion will be as indicated. We can represent this as:

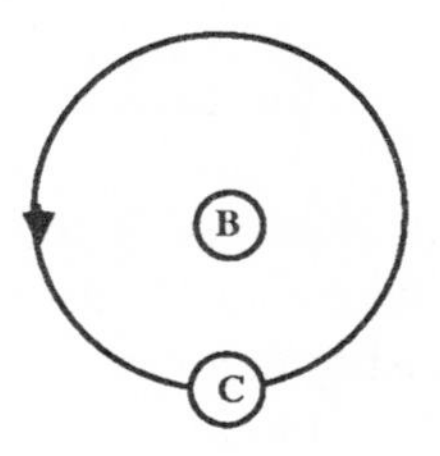


It is more difficult to determine how A's motion will be seen from B. It is clear, first of all, that the distance between A and B will vary from CA – CB to CA + CB. To represent C's motion, it will be necessary that it moves so that its distance from B will vary between these extremes. This requirement can be expressed mathematically by specifying that CA – CB ≤ AB ≤ CA + CB. As can be seen in the first diagram, CA and CB are fixed quantities, whereas AB varies. Let us construct a diagram that preserves this condition. First, draw a circle of radius CB with B at the center and with C moving on this circle. Draw another circle around B of radius CA. Clearly we cannot put C on this circle because then it would be at a constant distance from B, whereas its distance from B varies. Construct at some point D on the circle of radius AC a third circle of radius CB. This will ensure that the distance between A and B will always remain such that CA – CB ≤ AB ≤ CA + CB.

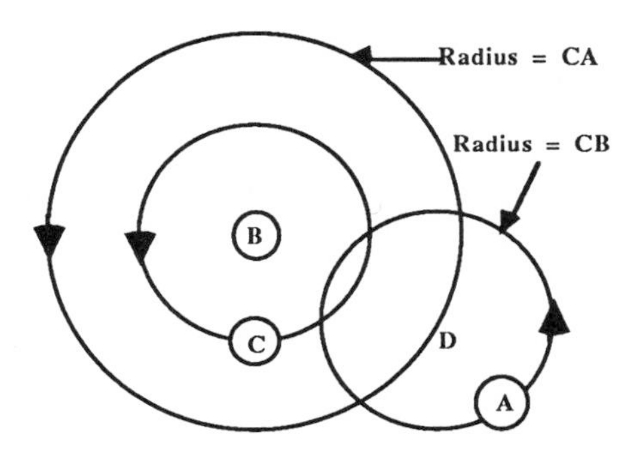


Two problems, however, come to mind. Because the two circles intersect, C and A can collide, whereas in our original diagram a collision between C and A would be impossible. Second, the above model does not maintain A at a constant distance from C, as is the case in the first diagram. How can the diagram be modified so that these problems are overcome? The next figure suggests a solution.

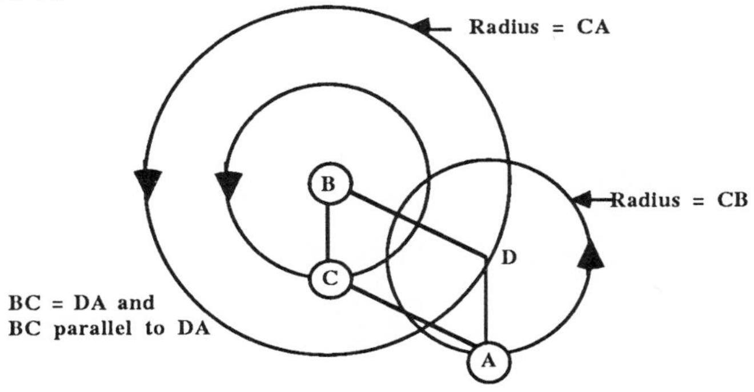

Suppose we require that the radius from D to A always remains parallel to the radius from B to C. This will allow AB to vary over the range $CA - CB \leq AB \leq CA + CB$, but it will preclude a collision and it will ensure that the distance between C and A remains fixed. The diagram indicates the directions of motion. The period of A as seen from B will be equal to $P_B$ because the circle on which it moves is linked to the motion of C. Careful inspection of this model makes evident that by means of it all the motions of C and A as seen from B can be represented.

# Chapter Two

## *Greek Astronomy before Ptolemy*

### *Introduction*

Among the achievements of the Greeks, few rival their development of astronomy. In studying the Greek contribution, it is well to keep in mind that other ancient civilizations also developed astronomical systems. The most impressive of these was created by the Babylonians, who formulated a system of which Ptolemy possessed some knowledge. The Babylonian approach, unlike the Greek, was primarily numerical. Although it merits careful study in its own right, we shall pass over it with only the note that it rivaled Greek astronomy in mathematical power.

### *Plato (428–348 B.C.)*

Although the famous philosopher Plato was not an astronomer, he frequently encouraged the study of that discipline in his writings. Among the reasons why he attributed importance to astronomy was the belief, mentioned in some of his writings, that the planets are divinities. His influence was not restricted to encouraging study of astronomy, which had already attained a significant level of sophistication by his time; he also made an influential proposal concerning the methodology of astronomy. In particular, as Simplicius commented, Plato set for astronomers the task of finding out "the uniform and ordered motions by the assumption of which the apparent movements of the planets can be accounted for." This has been interpreted to mean that astronomers should account for the planetary motions by means of combinations of perfect (regular) circular motions.

### *Eudoxus of Cnidus (ca. 400–ca. 347 B.C.)*

Among ancient astronomers and mathematicians, few surpassed Eudoxus in creativity. An associate of Plato's Academy, he devised an astronomical system, which, if the reconstruction of it that historians of astronomy have adopted is correct, ranks with the plays of the tragedians among the most impressive products of Greek genius. He explained the motions of the sun, moon, and planets by means of **concentric spheres,** the functioning of which can be illustrated by his model for the moon's motions.

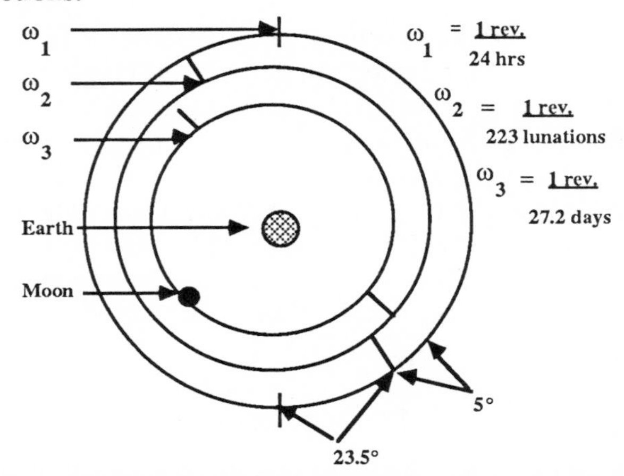


In the diagram, the outermost sphere should be viewed as rotating from east to west once each 24 hours on an axis passing through Polaris. Its motion accounts for the fact that the moon revolves around the earth once each day along with the planets and stars. Within this sphere and inclined to it at a 23 1/2° angle is a second sphere. It rotates in the same direction as the outermost sphere, but at a rate of one rotation every 223 lunations (a lunation is equivalent to one synodic month). The moon is on the third sphere, 90° down from the axis of rotation. The third sphere is inclined at 5° to the second sphere and rotates at the rate of one rotation every 27 1/5 days, turning in the direction opposite to that with which the first and second spheres rotate. It thus carries the moon through the zodiac. The 5° angle that its axis has to the second sphere accounts for the moon crossing the ecliptic. The middle sphere's motion accounts for the fact that the moon's draconitic period (the time between successive upward crossings of the ecliptic by the moon) is slightly less than its sidereal period. The three spheres should be conceived as connected and in simultaneous motion. When conceived in this manner, it becomes clear how this model was able to account effectively for the moon's chief motions.

The Eudoxian theory of planetary motions was more complex. In particular, he used 4 spheres for each planet, the additional sphere being necessary to produce planetary retrogressions. Because of this, the complete Eudoxian system involved 27 spheres, which were as follows:

| For the Motion of: | Number of Spheres: |
|---|---|
| Stars | 1 |
| Sun | 3 |
| Moon | 3 |
| Five Planets (four per planet) | 20 |
| **Total** | **27** |

Did Eudoxus believe his system to be physically true? Although only fragments of Eudoxus's astronomical writings have survived, various evidences point to the conclusion that he viewed it as only a mathematical system; the spheres function, in effect, as **computational devices** rather than as entities assumed to have physical existence. First, because all spheres are centered on the earth, the celestial bodies carried by them should remain at constant distances from the earth and thus not vary in brightness or apparent diameter. But it is obvious that the planets do vary in brightness; for example, the changes in brightness of Mars are so large that no observer can overlook them. Second, were these spheres physically real, the motions of the outermost spheres would be conveyed to the inner spheres, making it necessary to have counterturners that would undo the motions of the outer spheres. No such counterturners were attributed to the Eudoxian system. Third, none of the ancient accounts of the Eudoxian homocentric sphere system mentions Eudoxus having discussed the composition of his spheres. In short, Eudoxus seems to have viewed his system as only a mathematical model for representing the planetary motions.

*Callippus of Cyzicus (ca. 370–ca. 300 B.C.)*

Callippus adopted the Eudoxian system, but added additional spheres to it so as to enhance its accuracy. In particular, he added

2 additional spheres for both the sun and moon and 1 additional sphere for Mercury, Venus, and Mars. The total number of spheres in the Callippian system is consequently $27 + 2 + 2 + 1 + 1 + 1 = 34$.

*Aristotle (384–322 B.C.)*

Aristotle adopted the basic Eudoxian system as modified by Callippus, but changed it in a very important way. He made it **physical**; that is, he attributed real existence to the spheres, suggesting that they are composed of a perfect, transparent material. He believed that the motions of the heavens are produced by the Prime Mover, who acts from outside of the starry vault. Because of these ideas, it was necessary for him to postulate additional spheres, which act as **counterturners**, undoing the motions of the outer spheres. Were such counterturners not present, all the motions of the outermost planet would be passed on to the spheres next nearest the earth. Like Plato, Aristotle associated divinities with the planets. In summary:

| For the Motion of: | Number of Spheres: |
|---|---|
| Stars | 1 |
| Saturn | 4 |
| Counterturners | 3 |
| Jupiter | 4 |
| Counterturners | 3 |
| Mars | 5 |
| Counterturners | 4 |
| Venus | 5 |
| Counterturners | 4 |
| Mercury | 5 |
| Counterturners | 4 |
| Sun | 5 |
| Counterturners | 4 |
| Moon | 5 |
| **Total** | **56** |

In his *Physics*, Aristotle explained that the celestial motions must all be circular because only circular motions come back on themselves and can be eternal. Aristotle consequently sharply distinguished the realm of the eternal, incorruptible heavens from the terrestrial realm. To explain terrestrial motions, he urged that material objects move downward toward their **natural place,** which is the center of the earth or, in the case of air and fire, ascend upward, again toward their natural place.

Aristotle was not only well aware of the spherical shape of the earth, but also provided a number of arguments for its sphericity in his writings. For example, in his *On the Heavens*, he noted that the shadow of the earth in a lunar eclipse is circular and also that as we move to the south new constellations of stars become visible. Moreover, he stressed that, compared to the celestial region, the sphere of the earth is of small size.

### *Heraclides of Pontus (ca. 388–ca. 315 B.C.)*

Heraclides (not to be confused with Heraclitus) proposed that the earth rotates. Does previously provided information about the celestial motions suggest what might have led him to this claim? How is it possible to reconcile this idea with the daily rotation of the starry vault? It is sometimes stated, although no satisfactory evidence supports it, that Heraclides also suggested that Venus and Mercury orbit the sun. How would this idea fit with the fact that these two planets have bounded elongation?

### *Aristarchus of Samos (ca. 310–ca. 230 B.C.)*

Whereas the best known Greek tragedies were created during the fifth century B.C. and the finest Greek philosophic writings came from the fourth century B.C., Greek astronomy and mathematics continued to flourish in later periods. A striking example of this is Aristarchus, who in the third century B.C. made two very important contributions. The first of these can be read in a work by him that has been preserved; its title is: *On the Sizes and Distances of the Sun and Moon.* In that work, Aristarchus presented a method for determining the relative radii of the sun and moon and also the relative distances of those objects from us. The key observational element in his analysis of the **relative distances of the sun and moon** is that when the moon is in its half phase, the angle between the sun and moon is 87° (because the correct figure is 89° 50', the distance he determined is too small). Because the triangle EMS is right angled (see diagram), he could use trigonometric methods available to him to calculate the **relative** distances of the moon and sun from the earth.

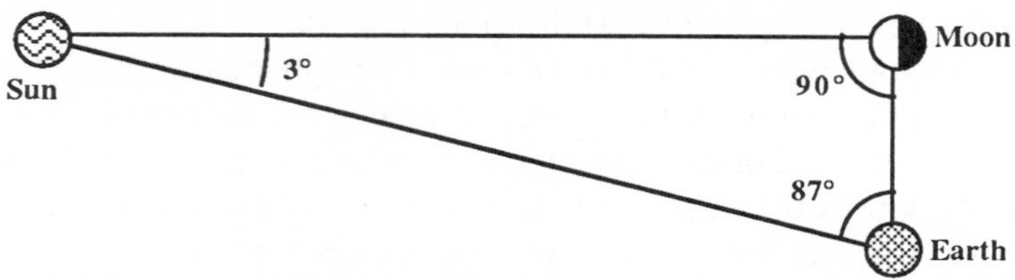


When he did this, he obtained the result that the sun is between 18 and 20 times farther from us than the moon (the sun is actually about 400 times farther from us than the moon). Earlier authors knew that the sun is farther from us than the moon, this being evident from the fact that during a solar eclipse the moon passes in front of the sun; nonetheless, previous values for their relative distances as attributed to Eudoxus and to Archimedes' father were about half the size of the figure Aristarchus attained. Aristarchus also derived values for the **sizes of the sun and moon** as compared to the size of the earth. His method for doing this depended on the observation that during a solar eclipse, the moon just covers the sun, indicating that their angular diameters as seen from the earth are essentially equal. He also had observational evidence from lunar eclipses that the shadow cone of the earth is, at the moon's distance, about twice as wide as the moon itself. This information permits the construction of the next diagram in which

$R_S$ = radius of the sun,
$R_E$ = the radius of the earth,
$R_M$ = the radius of the moon,
$D_M$ = distance of moon from earth, and
$D_S$ = distance of sun from earth.

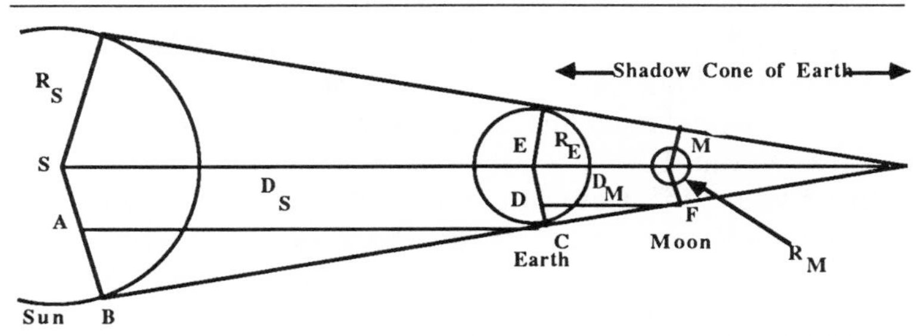


As can be seen by inspection of the diagram, triangles ABC and DCF are similar and also $D_S = AC$ and $D_M = DF$. Inspection of the diagram also shows that $AB = R_S - SA = R_S - EC = R_S - R_E$ and that $DC = R_E - ED = R_E - MF = R_E - 2R_M$, the last result being based on the determination, mentioned above, that the shadow cone of the earth is, at the moon's distance, about twice as wide as the moon itself. Because triangles ABC and DCF are similar, we have $D_M/D_S = DC/AB$. By substitution, we obtain the equation $D_M/D_S = (R_E - 2R_M)/(R_S - R_E)$. If we now take the earlier result that $D_S = 19\ D_M$ and the result, derived from the fact that the moon and sun have the same angular diameter, that $R_S = 19R_M$, and substitute these results into our equation, we get $1/19 = (R_E - 2R_M)/(19R_M - R_E)$. This equation can then be solved for the ratio of $R_M/R_E$. In particular, we get $R_M/R_E = 20/57$, meaning that the earth is about three times larger than the moon. And from this and our earlier figure that the sun's radius is about 19 times larger than the moon's, we find that the sun's radius must be more than six times larger than the earth's radius. The method used by Aristarchus was somewhat more complicated than this; his results were that $108/43 < R_E/R_M < 60/19$ and that $19/3 < R_S/R_E < 43/6$. His geometrical method was entirely sound; however, his observational base was imprecise. Not only is the angle in the first diagram not 87° but rather 89° 50', but also the width of the earth's shadow cone at the moon is closer to three than to two moon diameters. Thus the sun is actually about 400 times farther from us than the moon and its diameter is about 109 times greater than that of the earth. Nonetheless, Aristarchus had attained very important results as well as a method that needed only greater observational accuracy to produce precise values.

Aristarchus is famous for a second reason. Although no remaining writings by him testify to the fact, it is clear from passages in extant treatises by his near contemporary Archimedes that Aristarchus put forth the thesis that the earth rotates and also revolves around the sun, the latter being taken as the center of the cosmos. Because of this radical claim, Aristarchus is frequently referred to as **"the Copernicus of Antiquity."** It is unknown how Aristarchus came upon this bold idea, which according to the historian Plutarch led Cleanthes to urge that charges of impiety be brought against Aristarchus. Nonetheless, you may be able to make a very plausible conjecture about how he arrived at this thesis, which conjecture should also suggest a linkage of this idea with his treatise on the sizes and distances of the sun and moon.

*Eratosthenes of Alexandria (ca. 276–ca. 195 B.C.)*

As noted previously, Aristotle presented a number of arguments to show that the earth is spherical; moreover, he commented that because even a relatively short trip to the north or south reveals new stars, the earth cannot be of great size. A quantitative estimation of the size of the earth was provided by Eratosthenes, who was sometimes called "Beta" because although his achievements did not place him in the lead in any single discipline, he ranked second in a number of areas. According to the standard account, the method employed by Eratosthenes depended on the assumptions (1) that the sun is sufficiently distant that its rays can be treated as if they were parallel, (2) that Alexandria and Syene are separated by 5,000 stades (an ancient unit of distance that is believed by some scholars to be equivalent to 148.8 meters) and (3) that these cities lie on the same meridian (a meridian for any terrestrial location is the line running from the north pole through that location to the south pole).

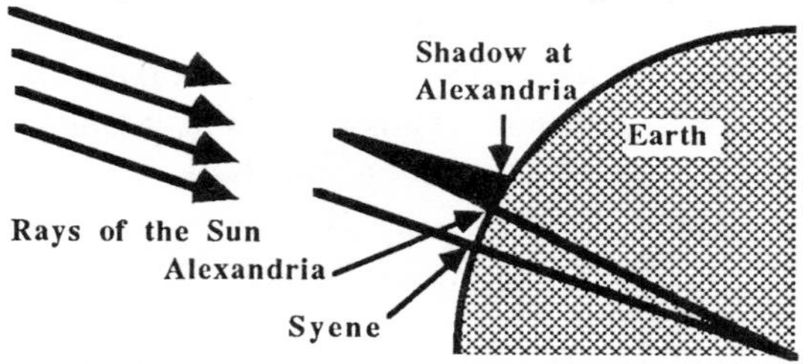


Basing his calculation on the observation that when the sun is directly overhead at Syene, a pole perpendicular to the earth at

Alexandria casts a shadow 1/50th of the length of the pole, Eratosthenes concluded that the distance from Alexandria to Syene must be 1/50th of the circumference of the earth, which correspondingly gives the circumference of the earth as (50x5000) = 250,000 stades. Converted to modern values, this is about 23,100 miles (modern value = ca. 24,900 miles).

*Apollonius of Perga (ca. 240–ca. 190 B.C.)*

Apollonius was the Greek mathematician who made the greatest contribution to the study of the conic sections: the ellipse, parabola, and hyperbola. He wrote a treatise in eight books on the conics, seven books of which have survived, the fifth, sixth, and seventh being known only in their Arabic translation. Apollonius also contributed to planetary theory; in particular, he is credited with having invented the **epicycle-deferent system**, which will be discussed shortly.

*Hipparchus of Nicaea (ca. 190–ca. 120 B.C.)*

Hipparchus is known not only as the most important Greek contributor to trigonometry, but also as an astronomer of exceptional talent. Unfortunately, almost all his astronomical writings have been lost, so that the nature and quality of his achievement must be assessed primarily on the basis of comments made by later Greek astronomers. Hipparchus made extensive use of the epicycle-deferent system and also of eccentric circles. We shall soon be discussing these in relation to his theories of the motions of the sun and moon.

In recent years, a major controversy has arisen concerning whether or not Ptolemy took many of his most important results from Hipparchus without adequately acknowledging his indebtedness. That Ptolemy owed much to Hipparchus is clear; what is disputed is the magnitude of that debt and Ptolemy's readiness to admit that he derived materials from Hipparchus. We need not enter into the details of this very heated controversy, which has attracted attention even in newspapers and popular magazines.

# Chapter Three

## *Some Mathematical Techniques of Ancient Astronomy*

By about 150 A.D., Greek astronomers possessed an array of powerful mathematical techniques to account for the celestial motions. Among these were the eccentric circle, the epicycle-deferent system, and the equant. Each of these devices is discussed in this chapter and illustrated by some of its elementary applications. This chapter also includes a discussion of Hipparchus's methods for predicting the positions of the sun and moon. Comprehension of these methods will set the stage for consideration in the next chapter of the most sophisticated astronomical system of antiquity, i.e., that presented by Ptolemy in his *Almagest*.

### *Eccentric Circle*

Imagine a point P revolving in a counterclockwise manner at a constant angular rate $\omega$ (omega) on a circle with center C and radius R. For this arrangement, an observer at C (the center of the circle) will see P move at a constant angular speed that will be $\omega = 360°/T$, where T (the time P takes to go through 360°) is the period of P.

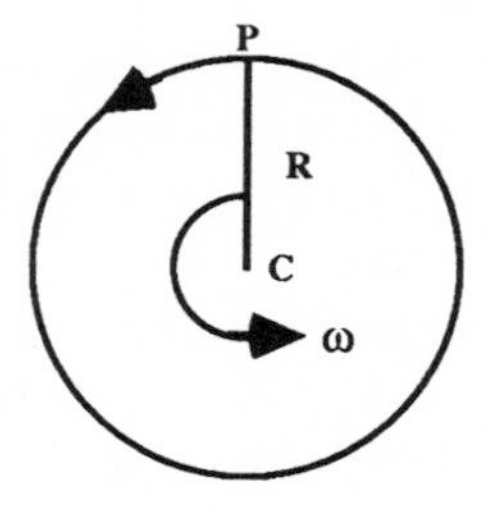


No celestial body moves in this simple fashion; for example, the sun and moon appear to speed up and to slow down somewhat in various parts of their orbits. Moreover, the planets at times appear even to retrogress. By introducing an **eccentric circle**, the Greeks were able to make a good first approximation to the motion of the sun. In an eccentric circle, the observer (O) on the earth is positioned at some distance e (the eccentricity) off the center (C) of the circle. The point P moves at a constant rate on the circle. As seen from O, however, P appears to slow down and to speed up.

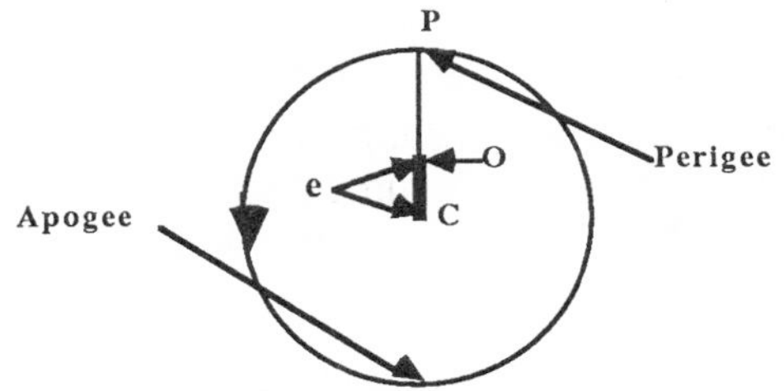


Where in its orbit will P appear to an observer at O to be moving most rapidly? Answer: at the top of its orbit, a point that is called the **perigee**. Where slowest? Answer: at the bottom of its orbit, at what is called its **apogee**.

*Illustration of the Use of an Eccentric Circle*

As noted previously, the eccentric circle can be used to account quite effectively for the motion of the sun. To see this, consult the next diagram in which P (the sun) moves at a constant angular speed on the circle centered at C. Suppose that its orbital period, T, is 365 days and that P moves in such a way that 89 days after leaving its perigee, it has moved 90°. After 182 1/2 days (89 days + 93 1/2 days or half its period), it has moved through 180°. After 276 days, it has progressed through 270°. Construct an eccentric circle that will account for these motions.

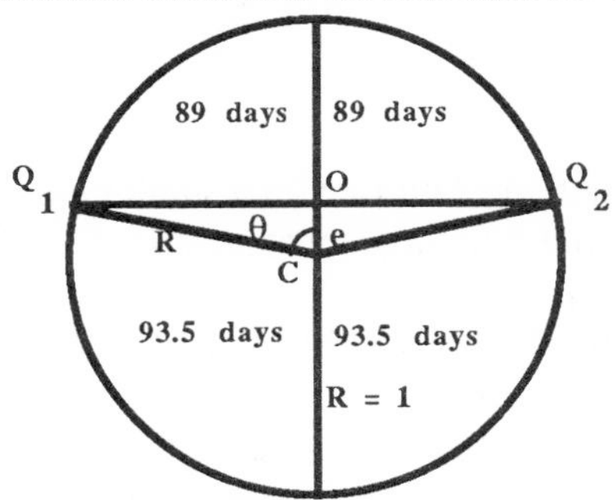


Solution: The diagram shows the points where P is seen by the observer at O as being 90° from perigee; these points, $Q_1$ and $Q_2$, are called the **quadrature** points. The eccentric circle is represented above; the difficulty is to find e in terms of R, which is the radius of the circle. Let us assume that R = 1 and find the value of e in relation to it. It is evident that $e = R\cos\theta$ and that $\theta/360° = 89^d/365^d$. Consequently we have $\theta = 360° \times 89/365 = 87° \ 49'$, from which it is apparent that $e = 1 \times \cos 87° \ 49' = .0381$. Although this case is simplified in a number of ways, it provides an idea of how the eccentric circle can be used to account for the motion of a point that appears to move at a non-constant speed, for example, the sun.

*Epicycle-Deferent System*

Another method, which is far more powerful than the eccentric circle, for representing the motion of a point moving around an observer is the epicycle-deferent system.

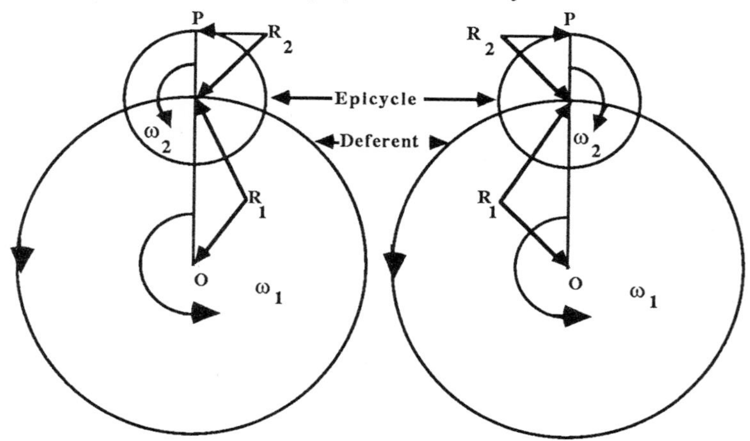


In it, P revolves at a constant rate on a small circle (the **epicycle**), the center of which moves at a constant (usually different) rate on another circle (the **deferent**). The observer is at the center of the larger circle. **Note**: By convention zero rate of revolution on the epicycle ($\omega_2 = 0$) is defined so that in this case $R_1$ and $R_2$ remain collinear. We could define zero rate of revolution as describing the case in which the epicycle radius continues to point straight up even as the center of the epicycle moves around the deferent circle, but shall not adopt this practice. Rather we shall conform to the definition of zero rate of revolution used by Ptolemy and the ancients. The center of the epicycle moves (by convention) in a counterclockwise direction, whereas the point P can move on the epicycle in either a counterclockwise direction (see diagram on the left) or a clockwise direction (diagram on right). The angle (see next diagram) between the deferent radius extended and the radius to the moving point is the **angle of anomaly** ($\alpha$). The angle between the deferent radius at the point at which it begins to move and its position at a given time is the **mean longitude** ($\lambda_m$) and the angle between the deferent radius in its initial position and the point (planet) at any given time is the **true longitude** ($\lambda$).

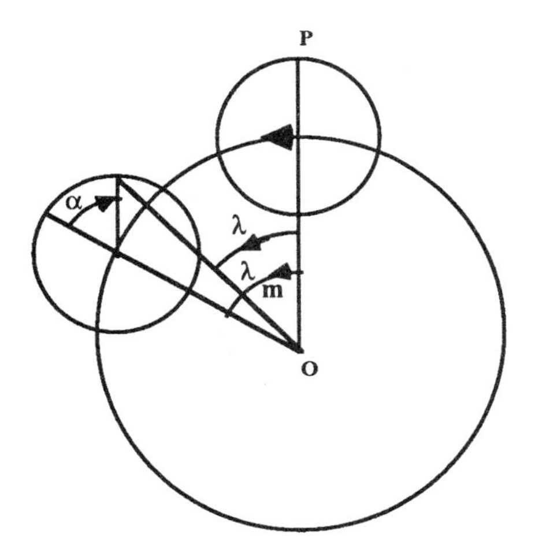


Some examples will illustrate the use of the epicycle-deferent system. Note that in each case the factors specified are the radius of the deferent, the radius of the epicycle, the rate of revolution of the center of the epicycle on the deferent, and the rate of revolution of the point P on the epicycle itself. Although these illustrations are presented in a geometrical form, it is important to keep in mind that this method provides a powerful technique for accounting for motions of celestial bodies. In the illustrations the celestial body will be represented by the point P, the motion of which will be traced. The illustrations, although relatively simple, should provide some understanding of how this technique can be applied to actual celestial motions.

**Illustration 1**: For $R_1 = 2$, $R_2 = 1$, $\omega_1$(counterclockwise) = 360°/yr, $\omega_2$(counterclockwise) = 0°/yr, plot the path of P.

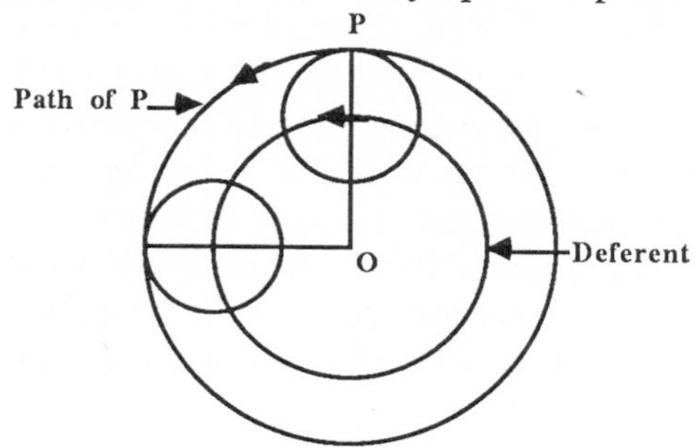


Solution: Because $\omega_2 = 0°$, the epicycle radius remains collinear with $R_1$. Consequently the path of P will be a circle of radius $R_1 + R_2$.

**Illustration 2**: Let $R_1 = 2$, $R_2 = 1$, $\omega_1$(counterclockwise) = 360°/yr, $\omega_2$(clockwise) = 360°/yr. Plot the path of P.

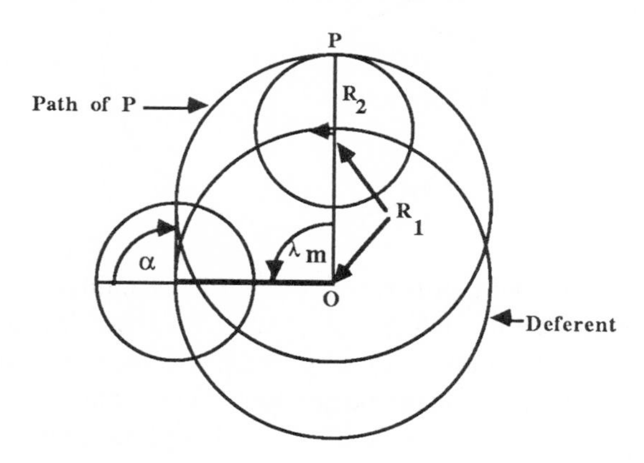


Solution: It is apparent that in this situation, the angle of anomaly will at every instant equal the angle of mean longitude. The diagram represents the case for a 90° angle of mean longitude; it is evident that the angle of anomaly at that instant is also 90°. The epicycle radius at that instant is pointing straight up. This, in fact, will be the case for every time value we select. Consequently, the resulting figure will be a circle of radius $R_1$ and with center located at a distance $R_2$ from O. This case is especially significant because it demonstrates that the epicycle-deferent system can be used to replicate any motion resulting from an eccentric circle. To achieve this, simply set the epicycle radius equal to the eccentricity of the eccentric circle.

**Illustration 3:** Let $R_1$ = 1.3 inches, $R_2$ = 0.5 inches, $\omega_1$(counterclockwise) = 1 rev./T, $\omega_2$(counterclockwise) = 3 rev./T. Determine the figure that results from these values.

**First Method:**
A convenient method for plotting epicyclic motion makes use of the next diagram, which is based on that given by Professors Barry M. Casper and Richard J. Noer in their *Revolutions in Physics* (New York, 1972).

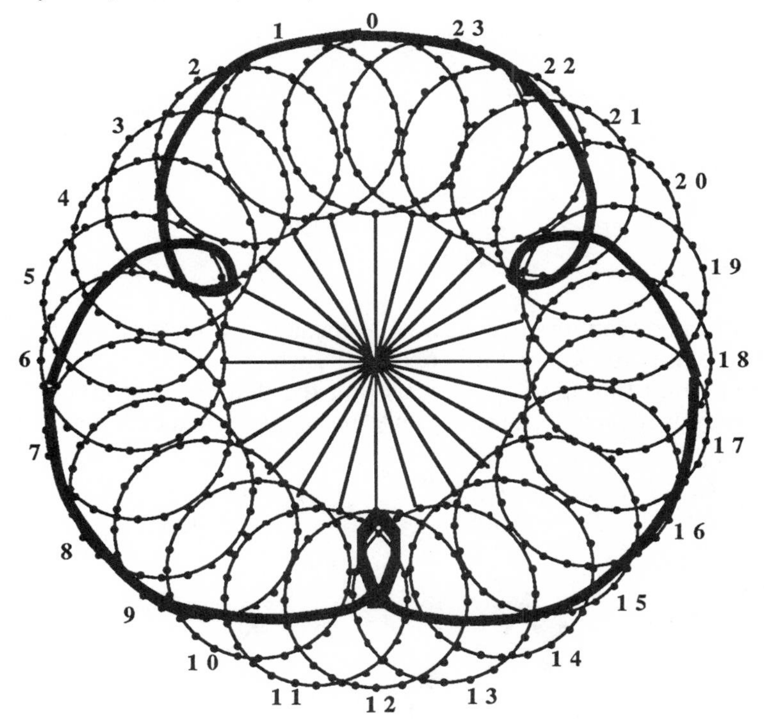


As represented in this diagram, the deferent and epicycle radii, $R_1$ and $R_2$, have values that conveniently correspond to the $R_1$ and $R_2$ values given in this problem. The diagram provides 24 epicycle positions, which correspond to angles of 15°, 30°, 45°, 60°, 75°, 90°, etc. We know that $\alpha$ equals $3\lambda_m$; hence for $\lambda_m = 15°$, $\alpha = 45°$, etc. Thus when the epicycle is in position 1, the point P will be on the third dot, i. e., the dot 45° around from the top point of the epicycle. When the epicycle is in position 2, the point P will be (3x30° =) 90° around from the top point of the epicycle. From such information, it is easy to plot the resulting path. This is shown by the dark line. An examination of this diagram raises questions as to where, when viewed from the center, does the point P move most rapidly? Where most slowly? Will there be any position at which P appears to move backwards?

**Second Method:**
Let us begin by calculating the values of $\alpha$ that correspond to a selected set of values of $\lambda_m$.

| $\lambda_m$ | $\alpha$ |
|---|---|
| 0° | 0° |
| 30° | 90° |
| 45° | 135° |
| 60° | 180° |
| 90° | 270° |
| 120° | 360° |

As can be seen, the $\alpha$ values will repeat as $\lambda_m$ increases above 120°; thus for $\lambda_m = 150°$, $\alpha = 450°$, which is equivalent to a revolution through 90°. Let us plot the deferent and epicycle positions separately, showing only the epicycles and deferents.

For $\lambda_m = 0°$, $\alpha = 0°$ | For $\lambda_m = 30°$, $\alpha = 90°$ | For $\lambda_m = 45°$, $\alpha = 135°$

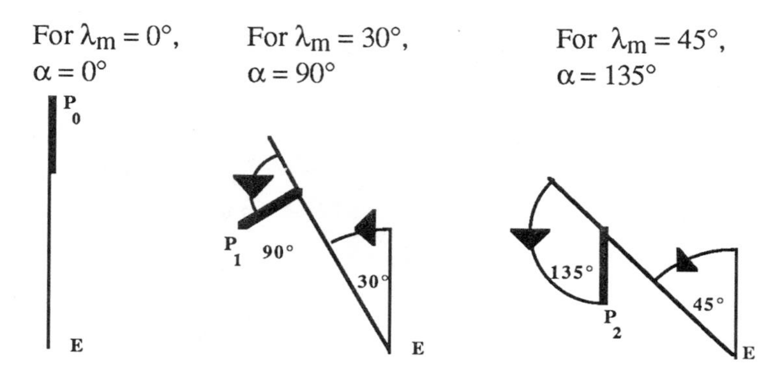

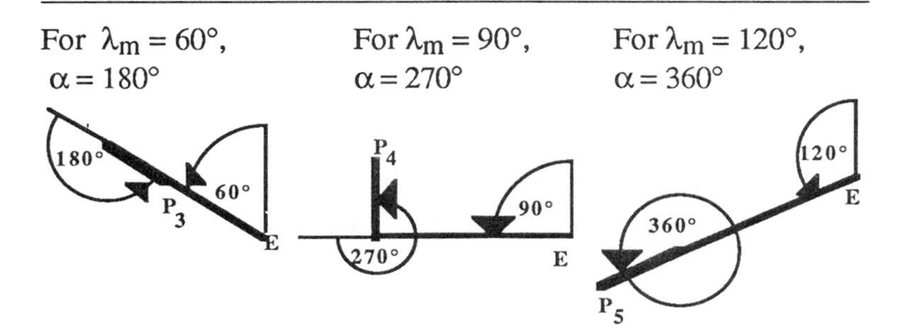


Let us now superimpose these six diagrams onto each other, taking the point E as the common point. As the next diagram shows, this will generate the same path as that obtained by the first method.

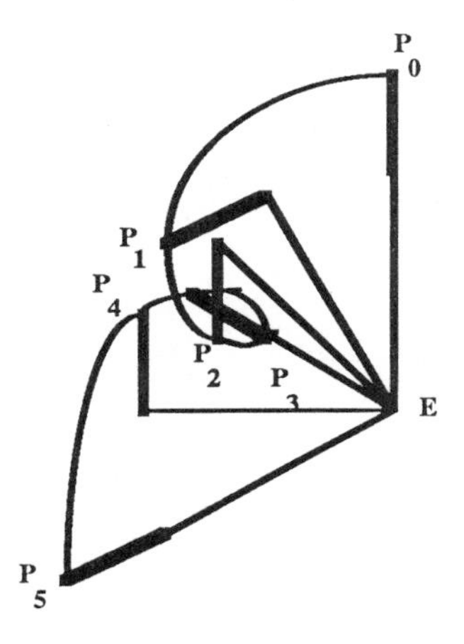


Although this is obviously a cumbersome method, it makes possible an interesting side result. Let us suppose that a person is positioned on the end of the deferent radius opposite E and suppose that this person is actually stationary. From that person's perspective, how will points P and E appear to move? We can determine this by again superimposing our six diagrams, taking in this case the end of the deferent as the one common point. From the next diagram, is it possible to infer the paths of P and E?

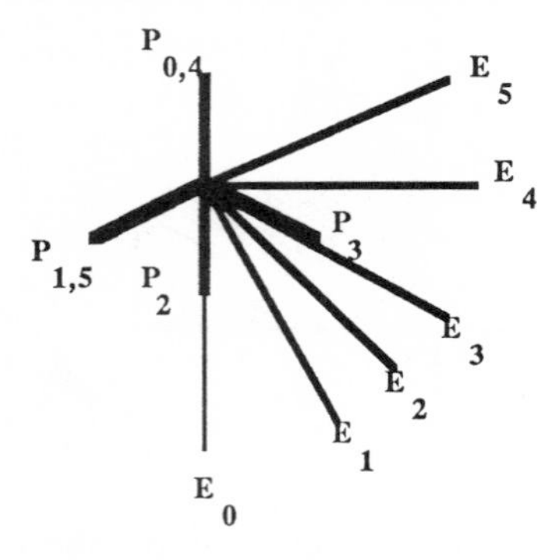


Does this suggest anything about why the epicycle-deferent system works so well in accounting for the motion of Mercury and Venus?

*The Equant*

In addition to the eccentric and the epicycle-deferent systems, Ptolemy, a later Greek astronomer, used a device called the **equant**.

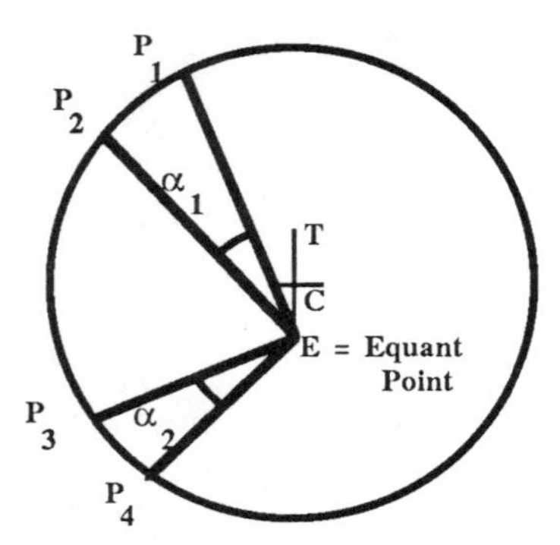


Let P be a point moving on a circle, C the center of the circle, T(terra) the location of the earth, and E the equant point. E is positioned on the extension of the line CT at a distance CE from C, and CE is set equal to CT. Ptolemy stipulated that the point P moves in such a manner that it moves through **equal angles around E in equal times.** Consequently, an observer on the earth will **not** see P move at a constant rate. In the figure, the angles $\alpha_1$ and $\alpha_2$ are equal; hence the time from $P_1$ to $P_2$ is equal to that from $P_3$ to $P_4$. It is important to ask two questions about the equant. First, in what section of the circle is P moving most rapidly and where most slowly? Second, does the equant point violate perfect circular motion?

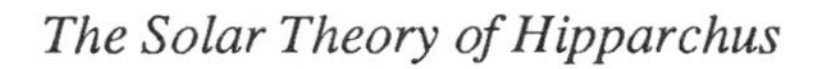

*The Solar Theory of Hipparchus*

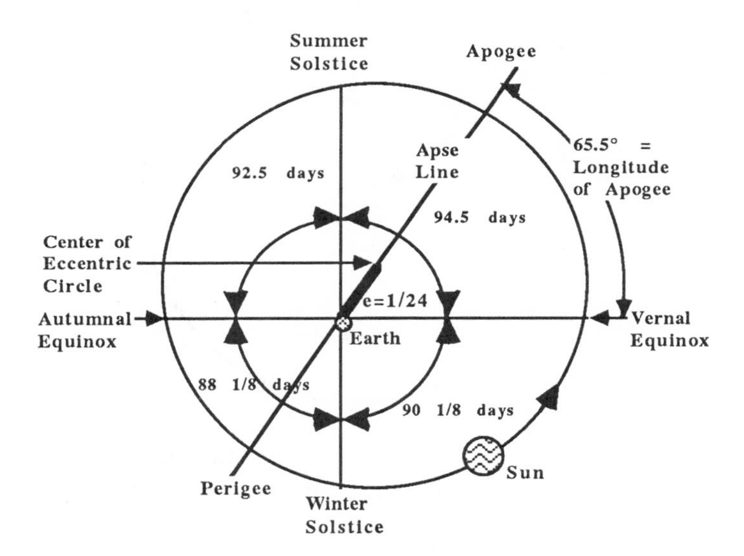


Understanding the solar theory of Hipparchus will provide help in comprehending the mathematical methods of Greek astronomy and also will set the stage for Ptolemy. Hipparchus knew that the sun takes about 94 1/2 days to move from the vernal equinox to the summer solstice. He was also aware that the sun moves from the summer solstice to the autumnal equinox in about 92 1/2 days. On the basis of these values, he devised an eccentric circle model to account for the solar motions. Because it takes so long for the sun to move from VE to SS, he saw that the center of the eccentric circle should be placed in the direction of the arc from VE to SS. This makes the sun move most slowly in this region. In particular, he was able to determine that an eccentric circle of eccentricity = 1/24 with its line of apse (the line from apogee to perigee) inclined at an angle of 65 1/2° around from VE toward SS would account fairly well for the motion of the sun. The sun moves on the circle at a rate of 360°/365.25days = 59'8" per day. Note: One degree consists of 60' (minutes) of arc; one minute of arc consists of 60"(seconds) of arc.

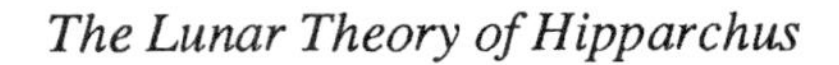

*The Lunar Theory of Hipparchus*

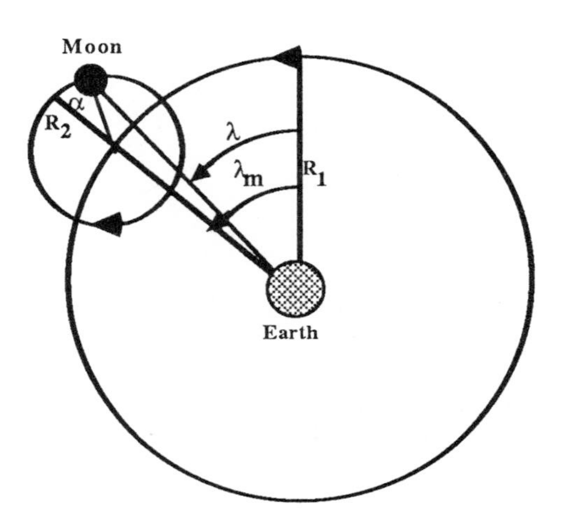


To account for the moon's motion, Hipparchus used an epicycle-deferent system (see diagram) with the moon revolving clockwise on the epicycle, while the center of the epicycle moves counterclockwise on the deferent. The epicycle is inclined at 5° to the plane of the deferent. In the figure, $\lambda_m$ is the mean longitude, $\alpha$ is the angle of anomaly, and $\lambda$ is the true longitude. The particular values Hipparchus used are $R_1 = 60$ and $R_2 = 5\ 1/4$. The epicycle center makes one complete revolution in one sidereal period of the moon (27.322 days). Thus $\omega_1$(counterclockwise) = $360°/27.322^d$ = 13°10'35" per day and $\omega_2$(clockwise) was set at 13°3'54" per day. This is necessary to reproduce the draconitic period of the moon. This system works well for the longitudes around 0° and 180°, but is less successful at the quadrature points. Note that in the diagram of the lunar theory of Hipparchus, the epicycle and deferent are not drawn to scale.

*Exercises on the Mathematical Methods of Greek Astronomy*

1. Using the diagram below, determine the figure that will result from the combination $\omega_1$(counterclockwise) = 1 rev./yr. and $\omega_2$ (counterclockwise) = 4 rev./yr.

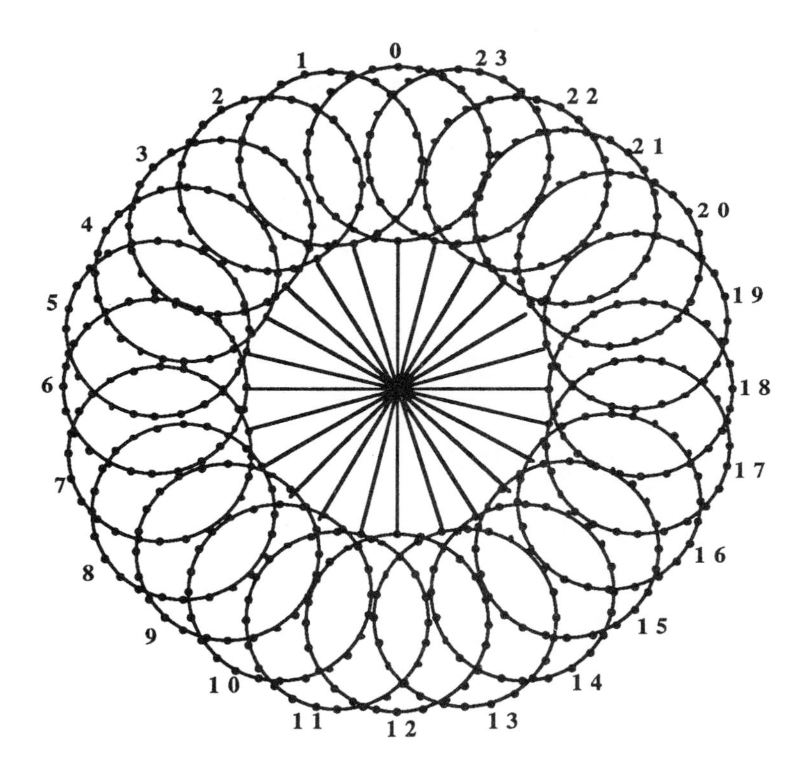


2. To represent the motion diagrammed below, what values for $\omega_1$ and $\omega_2$ should be used?

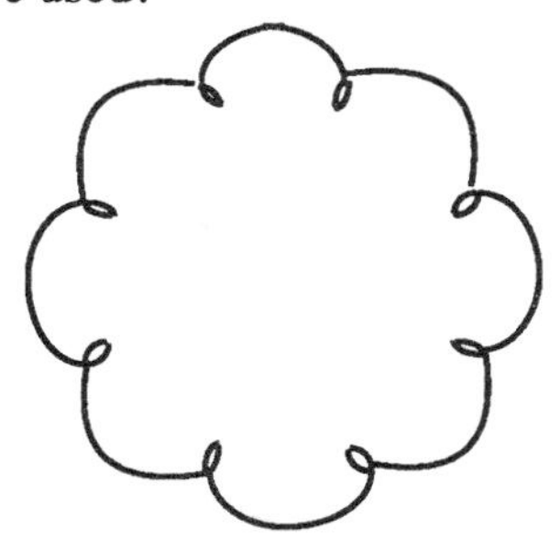

3. Using the diagram provided, determine the path that will result from $\omega_1$ (counterclockwise) = 1 rev. in 29 days and $\omega_2$(clockwise) = 1 rev. in 14.5 days.

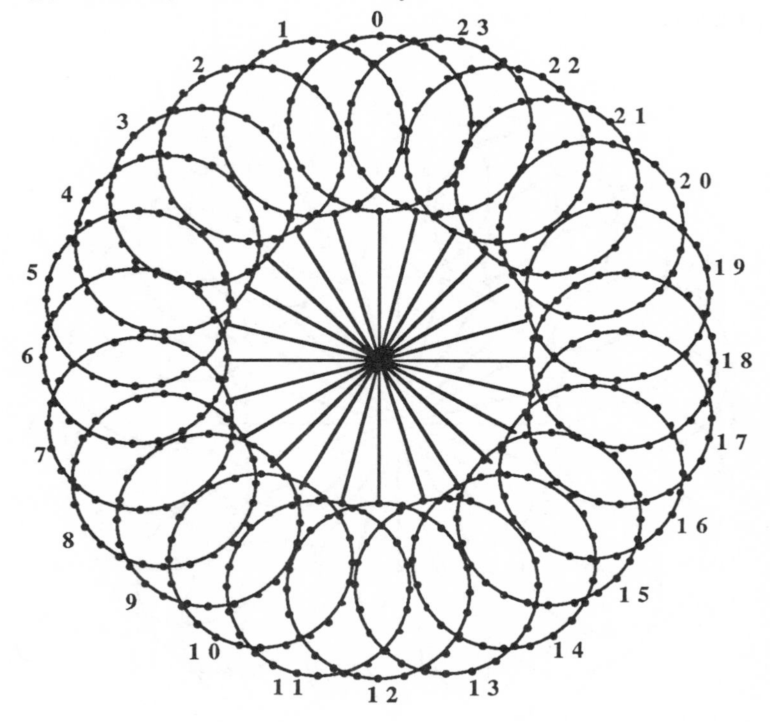

# Chapter Four

## *The Ptolemaic System*

### *Claudius Ptolemy of Alexandria (ca. 100–ca. 170 A.D.)*

Although Claudius Ptolemy (not to be confused with any of the kings of Egypt named Ptolemy) is recognized as having been among the most talented of the Greek scientists, little is known about his life. He lived in Alexandria, which after around 300 B.C. had become a leading center of Greek culture. He was no doubt associated with the Museum in Alexandria, an institution that included a school and great library of manuscripts. Ptolemy's most important astronomical treatise is his ***Almagest***, which word comes from the title that medieval Arabic astronomers used for his book. Its main Greek title was μαθηματικὴ σύνταξις (meaning "mathematical compilation"). In later Greek antiquity, it was at times referred to as μεγάλη σύνταξις or μεγίστη σύνταξις (the great, or greatest, compilation). According to one theory, the Arabic word "almagest" was derived from this. Thus there is some basis for referring to it as the "greatest book."

In his *Almagest*, which consists of thirteen books, Ptolemy presented his astronomical system in a quite highly mathematical manner. Filled with diagrams, formulae, and tables, it would take a few months or more to read. Translated into Arabic around 800 and into Latin in the late thirteenth century, it remained for over thirteen centuries as the premier presentation of mathematical astronomy. What Euclid's *Elements* was to geometry, Ptolemy's *Almagest* was to astronomy.

Ptolemy wrote a number of other works, the most important of which were completed after his *Almagest*. In astronomy, he compiled what is called his *Handy Tables*, and also wrote an important work named *Planetary Hypotheses*, in which he adopted a more physical approach to the planetary motions. For example, he attempted to compute the distances of the planets. This was not a topic treated in the *Almagest*, in which he had adopted the traditional ordering of the planets (moon, Mercury, Venus, sun, Mars, Jupiter, and Saturn) without attempting to assign distances to them. Other astronomical works by Ptolemy include his

*Analemma*, *Planisphaerium*, and *Phases of the Fixed Stars*. He also composed a work on astrology, the *Tetrabiblos*, which astrologers consider a classic. Among his other writings are his *Geography*, his *Harmonica* (a treatment of the mathematical theory of music), and a major and partly empirical investigation of optics, which includes a study of atmospheric refraction, a topic not treated in the *Almagest.*

*Ptolemy's Astronomical System: Sun and Moon*

Because of the complexity of Ptolemy's system, we shall not attempt to follow it in detail. One source of that complexity is that, in a sense, Ptolemy had not one, but rather a number of systems—one for each of the main bodies of our system. Of course, similarities exist among his models for the various bodies; it is on these that we shall focus our attention.

Ptolemy's treatment of the **sun** is essentially identical to that of Hipparchus. He used an eccentric circle to represent its motion, setting the eccentricity at 2 1/2 for a radius of 60. In other words, the orbit of the sun was centered on a point 1/24th of the radius of the sun's eccentric circle off from the position of the earth. This allowed him to account with reasonable accuracy for the motion of the sun. It is noteworthy that Ptolemy was not concerned in the *Almagest* with determining any orbital radii in absolute units. As will become evident, the length of the radius of every eccentric or deferent was set as 60 units. The other parameters, for example, epicycle radii, were then specified in terms of this standard length of 60.

Ptolemy's theory of the **moon** can be explained through the next four diagrams, which along with the main diagram for the planets have been derived in part from the excellent exposition and diagrams of the Ptolemaic system presented by F. S. Benjamin and G. J. Toomer in their *Campanus of Novara and Medieval Planetary Theory* (Madison, Wisconsin, 1971).

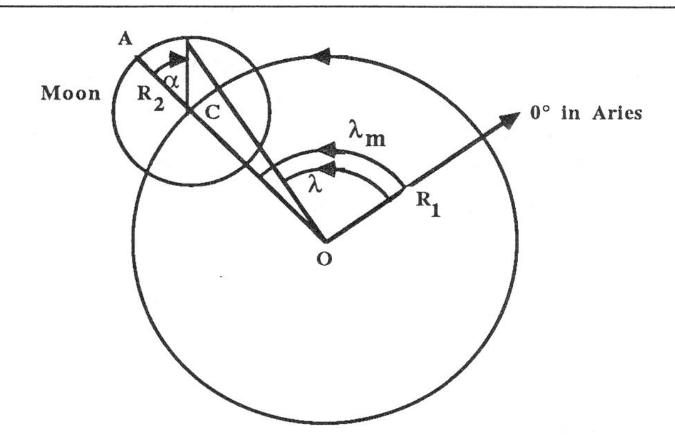


**Ptolemy's Simple Lunar Model**

The center C of the moon's epicycle moves on a deferent of radius $R_1 = 60$ from west to east with a uniform angular speed $\omega_1$(counterclockwise) = 13° 10' 35" per day around the deferent center O, which was set on the center of the earth. The moon moves on an epicycle of radius $R_2 = 5\ 1/4$ and centered on C with a uniform angular speed $\omega_2$(clockwise) = 13° 3' 54" per day. Because the rate of revolution of the moon on its epicycle is slower than the rate of revolution of the epicycle on the deferent, the moon's average angle of anomaly ($\alpha$) is less than its mean longitude ($\lambda_m$). In the above diagram, we imagine that when the moon was last at the apogee, A, then C was at 0° in the constellation Aries. The epicycle center C has moved through $\lambda_m$ while the moon has moved through $\alpha$ on the epicycle, giving it an actual longitude $\lambda$.

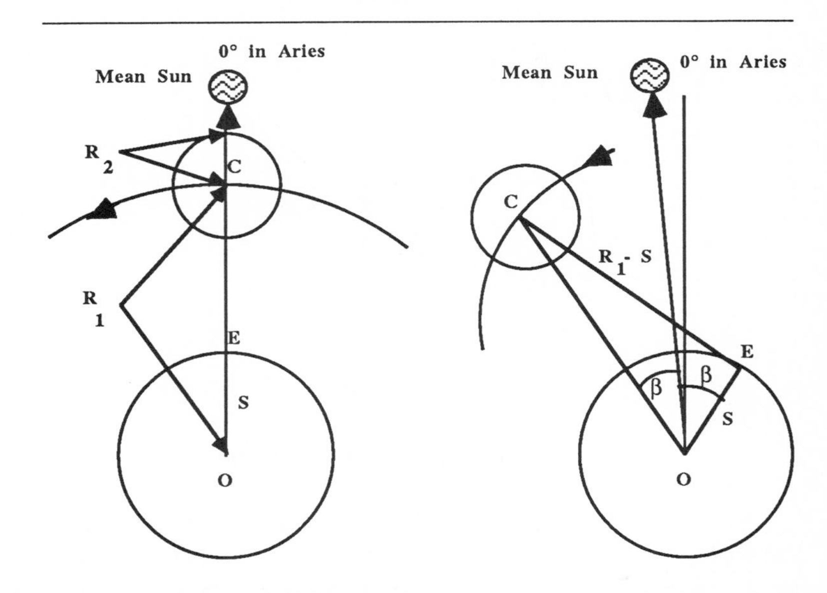


**Ptolemy's Refined Lunar Model: The "Crank" Mechanism**

Although fairly accurate for the moon near apogee or perigee, Ptolemy's simple lunar model gives poorer predictions when the moon is 90° from the sun, i.e., at the quadrature points. Thus Ptolemy devised a mechanism to increase the effect of the moon's epicycle by pulling the epicycle toward O (the earth). The radius $R_1$ of the moon's deferent is "hinged" at E, dividing it into $R_1 - S$ and S. As C moves counterclockwise from west to east around O, E moves clockwise on the small circle centered at O with radius S. E moves so that the angle β between E and the mean sun always equals the angle between the mean sun and epicycle center C. The figure at the left shows the situation at mean conjunction when the epicycle center has the same longitude as the mean sun. For simplicity, we have taken this as 0° in Aries. In this case, the "crank" is at full length. The figure at the right portrays the situation a few days later. Both C and the mean sun have moved to the east, while E has moved in the opposite direction, but through an angle equal to the angle separating C from the mean sun. The crank brings C closer to O, so that OC is now less than $R_1$. The maximum distance between the moon and earth on this model is $R_1 + R_2$, whereas the minimum distance is $R_1 - 2S - R_2$.

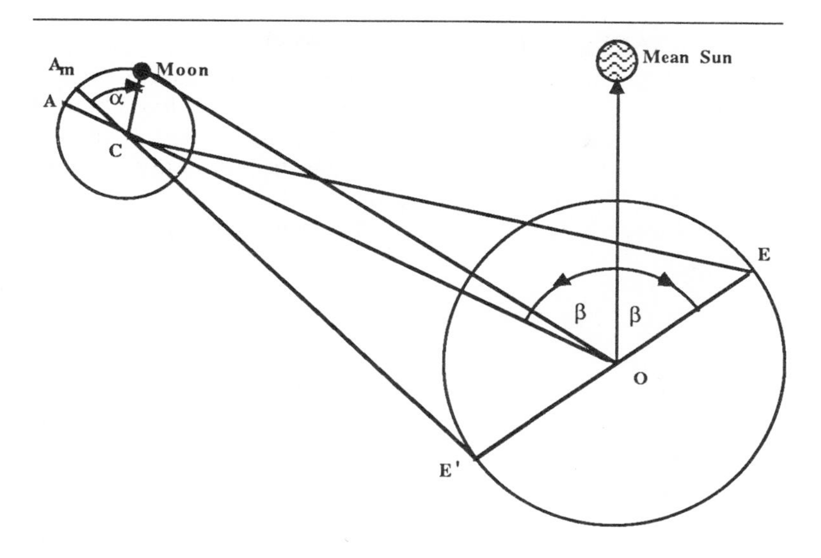


**Ptolemy's Fully Refined Lunar Model**

The epicycle center C moves at a constant rate in a counterclockwise direction around O. The point E moves in the opposite sense (i.e., clockwise) on the small circle about O (the earth), moving at a uniform rate with respect to the mean sun. This brings the epicycle gradually closer to O but then farther from it. The moon moves uniformly clockwise about the epicycle center C. Its mean anomaly $\alpha$ is measured from the "mean apogee" of the epicycle $A_m$, which is the point on the epicycle farthest from E´, the point on the small circle diametrically opposite E.

It is interesting and important to ask whether Ptolemy believed his model for the moon's motion to be physically true. The preceding information provides a basis for answering that question. As noted in the discussion of Ptolemy's refined lunar model, the distance of the moon ranges from a maximum of $R_1 + R_2$ to a minimum of $R_1 - 2S - R_2$. If the numerical values for these quantities are substituted for them, it becomes evident that the maximum value is 60 + 5 1/4 = 65 1/4, whereas the minimum is 60 –20(2/3) – 5 1/4 = about 34. This entails that the moon should at times appear nearly to double in size as it moves through its orbit. Ptolemy must have been aware that nothing of this kind happens; in fact, the angular width of the moon ranges between 29'30" and 33'36". This suggests that Ptolemy viewed his lunar models as **calculational devices**, rather than representations of the actual physical system. In other words, his eccentrics, epicycles, and deferents were seen by him as only hypothetical constructs for use in computation. Some later authors attributed reality to these devices, but such was not Ptolemy's view. Nonetheless, Ptolemy seems to have been fully convinced of the truth of the view that the earth is the stationary center of the cosmos.

*Ptolemy's System: The Planets*

The following diagram and discussion are focused primarily on Ptolemy's models for Mars, Jupiter, and Saturn. His models for Mercury and Venus are somewhat different and will not be treated in detail.

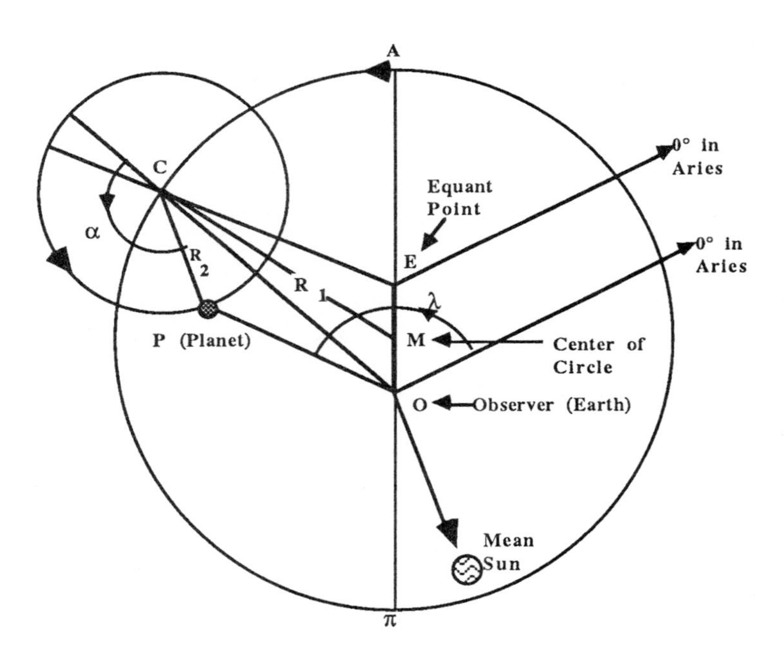


**Ptolemy's Model for Mars, Jupiter, or Saturn**

The epicycle's center C moves on the deferent of radius $R_1$, center M, and eccentricity from the earth (O) of MO. The uniform motion of the epicycle center takes place, not with reference to M or to O, but with reference to the **equant** point, E, which lies on the apse line AEMO$\pi$ at a distance from M equal to MO. The planet P moves on the epicycle of radius $R_2$ in the same sense as that in which C moves about M and at such a rate that **the radius CP always remains parallel to the line of direction from the earth O to the mean sun.**

The parameters that Ptolemy assigned for Mars, Jupiter, and Saturn are as follows:

| Planet | $R_1$ | $R_2$ | e(eccentricity) |
|---|---|---|---|
| Mars | 60 | 39;30 | 6;0 |
| Jupiter | 60 | 11;30 | 2;45 |
| Saturn | 60 | 6;30 | 3;25 |

Note that these values were all given in the sexagesimal system; 39;30, for example, should be read as 39 and 30 sixtieths. Note also that because Ptolemy gave all his $R_1$ values as a standard 60, these figures reveal nothing about how far those three planets were conceived to be from the earth. This is not a topic treated by Ptolemy in his *Almagest,* his chief concern having been to predict the zodiacal positions of the planets. The above table reveals that Ptolemy made the radius of the Martian epicycle quite large, whereas the epicyclic radii for Jupiter and Saturn were relatively smaller. It is very important to note Ptolemy's stipulation that the radius of the epicycle for each of those three planets must remain parallel to the line from the earth to the sun. What does this condition imply concerning when retrogradations occur; in particular, where relative to the earth and planet must the sun be for a retrogradation to occur? A study of the diagram may reveal an answer to this question. It is an empirical fact that the period between the successive retrogradations of a planet is equal to the planet's synodic period. Why on Ptolemy's model should this be so?

Another noteworthy feature of the Ptolemaic planetary system is its requirement that the centers of the Mercury and Venus epicycles must always be collinear with the sun. How does this ensure that bounded elongation will occur? It will prove interesting to compare Ptolemy's method with the Copernican explanation of the bounded elongations of Mercury and Venus. Ptolemy's parameters for Mercury and Venus are as follows:

| Planet | $R_1$ | $R_2$ | e(eccentricity) |
|---|---|---|---|
| Mercury | 60 | 22;30 | 3;0 |
| Venus | 60 | 43;10 | 3;0 |

It is interesting to ask whether any relation can be found between the epicycle radii of Mercury and Venus and the angular values for their bounded elongations.

A careful study of the Ptolemaic system reveals one very curious feature that it exhibits. This is that the motions of Mercury, Venus, Mars, Jupiter, and Saturn are all linked to the motion of the sun. In particular, as noted previously, the radii of Mars, Jupiter, and Saturn remain parallel to the line from the earth to the sun whereas the centers of the Mercury and Venus epicycles line up with the sun. Why should this linkage of all their motions to the sun be the case?

An additional problem treated by Ptolemy is that of the varying latitudes of the planets, i.e., their motions off the ecliptic. We shall pass over this. Even without a discussion of Ptolemy's impressive treatment of this problem, it should be clear why Ptolemy's *Almagest* was considered to be among the most impressive achievements of Greek thought.

The remainder of this chapter consists of the introductory sections of Book I of Ptolemy's *Almagest*. These eight sections contain the fundamental arguments presented by Ptolemy for his system as well as his objections to various other astronomical claims, for example, the idea that the earth rotates.

## CLAUDIUS PTOLEMY

# ALMAGEST[1]

### Book I

#### 1. {*Preface*}

The true philosophers, Syrus, were, I think, quite right to distinguish the theoretical part of philosophy, from the practical. For even if practical philosophy, before it is practical, turns out to be theoretical, nevertheless one can see that there is a great difference between the two: in the first place, it is possible for many people to possess some of the moral virtues even without being taught, whereas it is impossible to achieve theoretical understanding of the universe without instruction; furthermore, one derives most benefit in the first case [practical philosophy] from continuous practice in actual affairs, but in the other [theoretical philosophy] from making progress in the theory. Hence we thought it fitting to guide our actions (under the impulse of our actual ideas [of what is to be done]) in such a way as never to forget, even in ordinary affairs, to strive for a noble and disciplined disposition, but to devote most of our time to intellectual matters, in order to teach theories, which are so many and so beautiful, and especially those to which the epithet 'mathematical' is particularly applied. For Aristotle divides theoretical philosophy too, very fittingly, into three primary categories, physics, mathematics and theology. For everything that exists is composed of matter, form and motion; none of these [three] can be observed in its substratum by itself, without the others: they can only be imagined. Now the first cause of the first motion of the universe, if one considers it simply, can be thought of as an invisible and motionless deity; the division [of theoretical

---

[1]This reading consists of Chapters 1–8 of Book One of Ptolemy's *Almagest*. It is reprinted with the generous permission of Gerald Duckworth & Co., Ltd. from *Ptolemy's Almagest*, trans. and annotated by G. J. Toomer (New York: Springer-Verlag, 1984).

philosophy] concerned with investigating this [can be called] 'theology', since this kind of activity, somewhere up in the highest reaches of the universe, can only be imagined, and is completely separated from perceptible reality. The division [of theoretical philosophy] which investigates material and ever-moving nature, and which concerns itself with 'white', 'hot', 'sweet', 'soft' and suchlike qualities one may call 'physics'; such an order of being is situated (for the most part) amongst corruptible bodies and below the lunar sphere. That division [of theoretical philosophy] which determines the nature involved in forms and motion from place to place, and which serves to investigate shape, number, size, and place, time and suchlike, one may define as 'mathematics'. Its subject-matter falls as it were in the middle between the other two, since, firstly, it can be conceived of both with and without the aid of the senses, and, secondly, it is an attribute of all existing things without exception, both mortal and immortal: for those things which are perpetually changing in their inseparable form, it changes with them, while for eternal things which have an aethereal nature, it keeps their unchanging form unchanged.

From all this we concluded: that the first two divisions of theoretical philosophy should rather be called guesswork than knowledge, theology because of its completely invisible and ungraspable nature, physics because of the unstable and unclear nature of matter; hence there is no hope that philosophers will ever be agreed about them; and that only mathematics can provide sure and unshakeable knowledge to its devotees, provided one approaches it rigorously. For its kind of proof proceeds by indisputable methods, namely arithmetic and geometry. Hence we were drawn to the investigation of that part of theoretical philosophy, as far as we were able to the whole of it, but especially to the theory concerning divine and heavenly things. For that alone is devoted to the investigation of the eternally unchanging. For that reason it too can be eternal and unchanging (which is a proper attribute of knowledge) in its own domain, which is neither unclear nor disorderly. Furthermore it can work in the domains of the other [two divisions of theoretical philosophy] no less than they do. For this is the best science to help theology along its way, since it is the only one which can make a good guess at [the nature of] that activity which is unmoved and separated; [it can do this because] it is familiar with the attributes of those beings which are on the one hand perceptible, moving and being moved, but on the other hand eternal and unchanging, [I mean the attributes] having to do with motions and the arrangements of motions. As for physics, mathematics can make a significant contribution. For almost every peculiar attribute of material nature becomes apparent from the peculiarities of its motion from place to place. [Thus one can distinguish] the corruptible from the incorruptible by [whether it undergoes] motion in a straight line or in a circle, and heavy from light, and passive from active, by [whether it moves] towards the centre or away from the centre. With regard to virtuous conduct in practical actions and character, this science, above all things, could make men see clearly; from the constancy, order, symmetry and calm which are associated with the divine, it makes its followers lovers of this divine beauty, accustoming them and reforming their natures, as it were, to a similar spiritual state.

It is this love of the contemplation of the eternal and unchanging which we constantly strive to increase, by studying those parts of these sciences which have already been mastered by those who approached them in a genuine spirit of enquiry, and by ourselves attempting to contribute as much advancement as has been made possible by the additional time between those people and ourselves. We shall try to note down everything which we think we have discovered up to the present time; we shall do this as concisely as possible and in a manner which can be followed by those who have already made some progress in the field. For the sake of completeness in our treatment we shall set out everything useful for the theory of the heavens in the proper order, but to avoid undue length we shall merely recount what has been adequately established by the ancients. However, those topics which have not been dealt with [by our predecessors] at all, or not as usefully as they might have been, will be discussed at length, to the best of our ability.

## 2. {*On the order of the theorems*}

In the treatise which we propose, then, the first order of business is to grasp the relationship of the earth taken as a whole to the heavens taken as a whole. In the treatment of the individual aspects which follows, we must first discuss the position of the ecliptic and the regions of our part of the inhabited world and also the features differentiating each from the others due to the [varying] latitude at each horizon taken in order. For if the theory of these matters is treated first it will make examination of the rest easier. Secondly, we have to go through the motion of the sun and of the moon, and the phenomena accompanying these [motions]; for it would be impossible to examine the theory of the stars thoroughly without first having a grasp of these matters. Our final task in this way of approach is the theory of the stars. Here too it would be appropriate to deal first with the sphere of the so-called 'fixed stars', and follow that by treating the five 'planets', as they are called. We shall try to provide proofs in all of these topics by using as starting-points and foundations, as it were, for our search the obvious phenomena, and those observations made by the ancients and in our own times which are reliable. We shall attach the subsequent structure of ideas to this [foundation] by means of proofs using geometrical methods.

The general preliminary discussion covers the following topics: the heaven is spherical in shape, and moves as a sphere; the earth too is sensibly spherical in shape, when taken as a whole; in position it lies in the middle of the heavens very much like its centre; in size and distance it has the ratio of a point to the sphere of the fixed stars; and it has no motion from place to place. We shall briefly discuss each of these points for the sake of reminder.

### 3. {*That the heavens move like a sphere*}

It is plausible to suppose that the ancients got their first notions on these topics from the following kind of observations. They saw that the sun, moon and other stars were carried from east to west along circles which were always parallel to each other, that they began to rise up from below the earth itself, as it were, gradually got up high, then kept on going round in similar fashion and getting lower, until, falling to earth, so to speak, they vanished completely, then, after remaining invisible for some time, again rose afresh and set; and [they saw] that the periods of these [motions], and also the places of rising and setting, were, on the whole, fixed and the same.

What chiefly led them to the concept of a sphere was the revolution of the ever-visible stars, which was observed to be circular, and always taking place about one centre, the same [for all]. For by necessity that point became [for them] the pole of the heavenly sphere: those stars which were closer to it revolved on smaller circles, those that were farther away described circles ever greater in proportion to their distance, until one reaches the distance of the stars which become invisible. In the case of these, too, they saw that those near the ever-visible stars remained invisible for a short time, while those farther away remained invisible for a long time, again in proportion [to their distance]. The result was that in the beginning they got to the aforementioned notion solely from such considerations; but from then on, in their subsequent investigation, they found that everything else accorded with it, since absolutely all phenomena are in contradiction to the alternative notions which have been propounded.

For if one were to suppose that the stars' motion takes place in a straight line towards infinity, as some people have thought, what device could one conceive of which would cause each of them to appear to begin their motion from the same starting-point every day? How could the stars turn back if their motion is towards infinity? Or, if they did turn back, how could this not be obvious? [On such a hypothesis], they must gradually diminish in size until they disappear, whereas, on the contrary, they are seen to be greater at the very moment of their disappearance, at which time they are gradually obstructed and cut off, as it were, by the earth's surface.

But to suppose that they are kindled as they rise out of the earth and are extinguished again as they fall to earth is a completely absurd hypothesis. For even if we were to concede that the strict order in their size and number, their intervals, positions and periods could be restored by such a random and chance process; that one whole area of the earth has a kindling nature, and another an extinguishing one, or rather that the same part [of the earth] kindles for one set of observers and extinguishes for another set; and that the same stars are already kindled or extinguished for some observers while they are not yet for others: even if, I say, we were to concede all these ridiculous consequences, what could we say about the ever-visible stars, which neither rise nor set? Those stars which are kindled and extinguished ought to rise and set for observers everywhere, while those which are not kindled and extinguished ought always to be visible for observers everywhere. What cause could we assign for the fact that this is not so? We will surely not say that stars which are kindled and extinguished for some observers never undergo this process for other observers. Yet it is utterly obvious that the same stars rise and set in certain regions [of the earth] and do neither at others.

To sum up, if one assumes any motion whatever, except spherical, for the heavenly bodies, it necessarily follows that their distances, measured from the earth upwards, must vary, wherever and however one supposes the earth itself to be situated. Hence the sizes and mutual distances of the stars must appear to vary for the same observers during the course of each revolution, since at one time they must be at a greater distance, at another at a lesser. Yet we see that no such variation occurs. For the apparent increase in their sizes at the horizons is caused, not by a decrease in their distances, but by the exhalations of moisture surrounding the earth being interposed between the place from which we observe and the heavenly bodies, just as objects placed in water appear bigger than they are, and the lower they sink, the bigger they appear.

The following considerations also lead us to the concept of the sphericity of the heavens. No other hypothesis but this can explain how sundial constructions produce correct results; furthermore, the motion of the heavenly bodies is the most unhampered and free of all motions, and freest motion belongs among plane figures to the circle and among solid shapes to the sphere; similarly, since of different shapes having an equal boundary those with more angles are greater [in area or volume], the circle is greater than [all other] surfaces, and the sphere greater than [all other] solids; [likewise] the heavens are greater than all other bodies.

Furthermore, one can reach this kind of notion from certain physical considerations. E.g., the aether is, of all bodies, the one with constituent parts which are finest and most like each other; now bodies with parts like each other have surfaces with parts like each other; but the only surfaces with parts like each other are the circular, among planes, and the spherical, among three-dimensional surfaces. And since the aether is not plane, but three-dimensional, it follows that it is spherical in shape. Similarly, nature formed all earthly and corruptible bodies out of shapes which are round but of unlike parts, but all aethereal and divine bodies out of shapes which are of like parts and spherical. For if they were flat or shaped like a discus they would not always display a circular shape to all those observing them simultaneously from different places on earth. For this reason it is plausible that the aether surrounding them, too, being of the same nature, is spherical, and because of the likeness of its parts moves in a circular and uniform fashion.

### 4. {*That the earth too, taken as a whole, is sensibly spherical*}

That the earth, too, taken as a whole, is sensibly spherical can best be grasped from the following considerations. We can see, again, that the sun, moon and other stars do not rise and set simultaneously for everyone on earth, but do so earlier for those more towards the east, later for those towards the west. For we find that the phenomena at eclipses, especially lunar eclipses, which take place at the same time [for all observers], are nevertheless not recorded as occurring at the same hour (that is at an equal distance from noon) by all observers. Rather, the hour recorded by the more easterly observers is always later than that recorded by the more westerly. We find that the differences in the hour are proportional to the distances between the places [of observation]. Hence one can reasonably conclude that the earth's surface is spherical, because its evenly curving surface (for so it is when considered as a whole) cuts off [the heavenly bodies] for each set of observers in turn in a regular fashion.

If the earth's shape were any other, this would not happen, as one can see from the following arguments. If it were concave, the stars would be seen rising first by those more towards the west; if it were plane, they would rise and set simultaneously for everyone on earth; if it were triangular or square or any other polygonal shape, by a similar argument, they would rise and set simultaneously for all those living on the same plane surface. Yet it is apparent that nothing like this takes place. Nor could it be cylindrical, with the curved surface in the east-west direction, and the flat sides towards the poles of the universe, which some might suppose more plausible. This is clear from the following: for those living on the curved surface none of the stars would be ever-visible, but either all stars would rise and set for all observers, or the same stars, for an equal [celestial] distance from each of the poles, would always be invisible for all observers. In fact, the further we travel toward the north, the more of the southern stars disappear and the more of the northern stars appear. Hence it is clear that here too the curvature of the earth cuts off [the heavenly bodies] in a regular fashion in a north-south direction, and proves the sphericity [of the earth] in all directions.

There is the further consideration that if we sail towards mountains of elevated places from and to any direction whatever, they are observed to increase gradually in size as if rising up from the sea itself in which they had previously been submerged: this is due to the curvature of the surface of the water.

### 5. {*That the earth is in the middle of the heavens*}

Once one has grasped this, if one next considers the position of the earth, one will find that the phenomena associated with it could take place only if we assume that it is in the middle of the heavens, like the centre of a sphere. For if this were not the case, the earth would have to be either

[a] not on the axis [of the universe] but equidistant from both poles, or
[b] on the axis but removed towards one of the poles, or
[c] neither on the axis nor equidistant from both poles.

Against the first of these three positions militate the following arguments. If we imagined [the earth] removed towards the zenith or the nadir of some observer, then, if he were at *sphaera recta*, he would never experience equinox, since the horizon would always divide the heavens into two unequal parts, one above and one below the earth; if he were at *sphaera obliqua*, either, again, equinox would never occur at all, or, [if it did occur,] it would not be at a position halfway between summer and winter solstices, since these intervals would necessarily be unequal, because the equator, which is the greatest of all parallel circles drawn about the poles of the [daily] motion, would no longer be bisected by the horizon; instead [the horizon would bisect] one of the circles parallel to the equator, either to the north or to the south of it. Yet absolutely everyone agrees that these intervals are equal everywhere on earth, since [everywhere] the increment of the longest day over the equinoctial day at the summer solstice is equal to the decrement of the shortest day from the equinoctial day at the winter solstice. But if, on the other hand, we imagined the displacement to be towards the east or west of some observer, he would find that the sizes and distances of the stars would not remain constant and unchanged at eastern and western horizons, and that the time-interval from rising to culmination would not be equal to the interval from culmination to setting. This is obviously completely in disaccord with the phenomena.

Against the second position, in which the earth is imagined to lie on the axis removed towards one of the poles, one can make the following objections. If this were so, the plane of the horizon would divide the heavens into a part above the earth and a part below the earth which are unequal and always different for different latitudes, whether one considers the relationship of the same part at two different latitudes or the two parts at the same latitude. Only at *sphaera recta* could the horizon bisect the sphere; at a *sphaera obliqua* situation such that the nearer pole were the ever-visible one, the horizon would always make the part above the earth lesser and the part below the earth greater; hence another phenomenon would be that the great circle of the ecliptic would be divided into unequal parts by the plane of the horizon. Yet it is apparent that this is by no means so. Instead, six zodiacal signs are visible above the earth at all times and places, while the remaining six are invisible; then again [at a later time] the latter are visible in their entirety above the earth, while at the same time the others are not visible. Hence it is obvious that the horizon bisects the zodiac, since the same semi-circles are cut off by it, so as to appear at one time completely above the earth, and at another [completely] below it.

And in general, if the earth were not situated exactly below the [celestial] equator, but were removed towards the north or south in the direction of one of the poles, the result would be that at the equinoxes the shadow of the gnomon at sunrise would no longer form a straight line with its shadow at sunset in a plane parallel to the horizon, not even sensibly. Yet this is a phenomenon which is plainly observed everywhere.

It is immediately clear that the third position enumerated is likewise impossible, since the sorts of objection which we made to the first [two] will both arise in that case.

To sum up, if the earth did not lie in the middle [of the universe], the whole order of things which we observe in the increase and decrease of the length of daylight would be fundamentally upset. Furthermore, eclipses of the moon would not be restricted to situations where the moon is diametrically opposite the sun (whatever part of the heaven [the luminaries are in]), since the earth would often come between them when they were not diametrically opposite, but at intervals of less than a semi-circle.

### 6. {*That the earth has the ratio of a point to the heavens*}

Moreover, the earth has, to the senses, the ratio of a point to the distance of the sphere of the so-called fixed stars. A strong indication of this is the fact that the sizes and distances of the stars, at any given time, appear equal and the same from all parts of the earth everywhere, as observations of the same [celestial] objects from different latitudes are found to have not the least discrepancy from each other. One must also consider the fact that gnomons set up in any part of the earth whatever, and likewise the centres of armillary spheres, operate like the real centre of the earth; that is, the lines of sight [to heavenly bodies] and the paths of shadows caused by them agree as closely with the [mathematical] hypotheses explaining the phenomena as if they actually passed through the real centre-point of the earth.

Another clear indication that this is so is that the planes drawn through the observer's lines of sight at any point [on earth], which we call 'horizons', always bisect the whole heavenly sphere. This would not happen if the earth were of perceptible size in relation to the distance of the heavenly bodies; in that case only the plane drawn through the centre of the earth could bisect the sphere, while a plane through any point on the surface of the earth would always make the section [of the heavens] below the earth greater than the section above it.

### 7. {*That the earth does not have any motion from place to place, either*}

One can show by the same arguments as the preceding that the earth cannot have any motion in the aforementioned directions, or indeed ever move at all from its position at the centre. For the same phenomena would result as would if it had any position other than the central one. Hence I think it is idle to seek for causes for the motion of objects towards the centre, once it has been so clearly established from the actual phenomena that the earth occupies the middle place in the universe, and that all heavy objects are carried towards the earth. The following fact alone would most readily lead one to this notion [that all objects fall towards the centre]. In absolutely all parts of the earth, which, as we said, has been shown to be spherical and in the middle of the universe, the direction and path of the motion (I mean the proper, [natural] motion) of all bodies possessing weight is always and everywhere at right angles to the rigid plane drawn tangent to the point of impact. It is clear from this fact that, if [these falling objects] were not arrested by the surface of the earth, they would certainly reach the centre of the earth itself, since the straight line to the centre is also always at right angles to the plane tangent to the sphere at the point of intersection [of that radius] and the tangent.

Those who think it paradoxical that the earth, having such a great weight, is not supported by anything and yet does not move, seem to me to be making the mistake of judging on the basis of their own experience instead of taking into account the peculiar nature of the universe. They would not, I think, consider such a thing strange once they realised that this great bulk of the earth, when compared with the whole surrounding mass [of the universe], has the ratio of a point to it. For when one looks at it in that way, it will seem quite possible that that which is relatively smallest should be overpowered and pressed in equally from all directions to a position of equilibrium by that which is the greatest of all and of uniform nature. For there is no up and down in the universe with respect to itself, any more than one could imagine such a thing in a sphere: instead the proper and natural motion of the compound bodies in it is as follows: light and rarefied bodies drift outwards towards the circumference, but seem to move in the direction which is 'up' for each observer, since the overhead direction for all of us, which is also called 'up', points towards the surrounding surface; heavy and dense bodies, on the other hand, are carried towards the middle and the centre, but seem to fall downwards, because, again, the direction which is for all of us towards our feet, called 'down', also points towards the centre of the earth. These heavy bodies, as one would expect, settle about the centre because of their mutual pressure and resistance, which is equal and uniform from all directions. Hence, too, one can see that it is plausible that the earth, since its total mass is so great compared with the bodies which fall towards it, can remain motionless under the impact of these very small weights (for they strike it from all sides), and receive, as it were, the objects falling on it. If the earth had a single motion in common with other heavy objects, it is obvious that it would be carried down faster than all of them because of its much greater size: living things and individual heavy objects would be left behind, riding on the air, and the earth itself would very soon have fallen completely out of the heavens. But such things are utterly ridiculous merely to think of.

But certain people, [propounding] what they consider a more persuasive view, agree with the above, since they have no argument to bring against it, but think that there could be no evidence to oppose their view if, for instance, they supposed the heavens to remain motionless, and the earth to revolve from west to east about the same axis [as the heavens], making approximately one revolution each day; or if they made both heaven and earth move by any amount whatever, provided, as we said, it is about the same axis, and in such a way as to preserve the overtaking of one by the other. However, they do not realise that, although there is perhaps nothing in the celestial phenomena which would count against that hypothesis, at least from simpler considerations, nevertheless from what would occur here on earth and in the air, one can see that such a notion is quite ridiculous. Let us concede to them [for the sake of argument] that such an unnatural thing could happen as that the most rare and light of matter should either not move at all or should move in a way no different from that of matter with the opposite nature (although things in the air, which are less rare [than the heavens] so obviously move with a more rapid motion than any earthy object); [let us concede that] the densest and heaviest objects have a proper motion of the quick and uniform kind which they suppose (although, again, as all agree, earthy objects are sometimes not readily moved even by an external force). Nevertheless, they would have to admit that the revolving motion of the earth must be the most violent of all motions associated with it, seeing that it makes one revolution in such a short time; the results would be that all objects not actually standing on the earth would appear to have the same motion, opposite to that of the earth: neither clouds nor other flying or thrown objects would ever be seen moving towards the east, since the earth's motion towards the east would always outrun and overtake them, so that all other objects would seem to move in the direction of the west and the rear. But if they said that the air is carried around in the same direction and with the same speed as the earth, the compound objects in the air would none the less always seem to be left behind by the motion of both [earth and air]; or if those objects too were carried around, fused, as it were, to the air, then they would never appear to have any motion either in advance or rearwards: they would always appear still, neither wandering about nor changing position, whether they were flying or thrown objects. Yet we quite plainly see that they do undergo all these kinds of motion, in such a way that they are not even slowed down or speeded up at all by any motion of the earth.

## 8. {*That there are two different primary motions in the heavens*}

It was necessary to treat the above hypothesis first as an introduction to the discussion of particular topics and what follows after. The above summary outline of them will suffice, since they will be completely confirmed and further proven by the agreement with the phenomena of the theories which we shall demonstrate in the following sections. In addition to these hypotheses, it is proper, as a further preliminary, to introduce the following general notion, that there are two different primary motions in the heavens. One of them is that which carries everything from east to west: it rotates them with an unchanging and uniform motion along circles parallel to each other, described, as is obvious, about the poles of this sphere which rotates everything uniformly. The greatest of these circles is called the 'equator', because it is the only [such parallel circle] which is always bisected by the horizon (which is a great circle), and because the revolution which the sun makes when located on it produces equinox everywhere, to the senses. The other motion is that by which the spheres of the stars perform movements in the opposite sense to the first motion, about another pair of poles, which are different from those of the first rotation. We suppose that this is so because of the following considerations. When we observe for the space of any given single day, all heavenly objects whatever are seen, as far as the senses can determine, to rise, culminate and set at places which are analogous and lie on circles parallel to the equator; this is characteristic of the first motion. But when we observe continuously without interruption over an interval of time, it is apparent that while the other stars retain their mutual distances and (for a long time) the particular characteristics arising from the positions they occupy as a result of the first motion, the sun, the moon and the planets have certain special motions which are indeed complicated and different from each other, but are all, to characterise their general direction, towards the east and opposite to [the motion of] those stars which preserve their mutual distances and are, as it were, revolving on one sphere.

Now if this motion of the planets too took place along circles parallel to the equator, that is, about the poles which produce the first kind of revolution, it would be sufficient to assign a single kind of revolution to all alike, analogous to the first. For in that case it would have seemed plausible that the movements which they undergo are caused by various retardations, and not by a motion in the opposite direction. But as it is, in addition to their movement towards the east, they are seen to deviate continuously to the north and south [of the equator]. Moreover the amount of this deviation cannot be explained as the result of a uniformly-acting force pushing them to the side: from that point of view it is irregular, but it is regular if considered as the result of [motion on] a circle inclined to the equator. Hence we get the concept of such a circle, which is one and the same for all planets, and particular to them. It is precisely defined and, so to speak, drawn by the motion of the sun, but it is also travelled by the moon and the planets, which always move in its vicinity, and do not randomly pass outside a zone on either side of it which is determined for each body. Now since this too is shown to be a great circle, since the sun goes to the north and south of the equator by an equal amount, and since, as we said, the eastward motion of all of the planets takes place on one and the same circle, it became necessary to suppose that this second, different motion of the whole takes place about the poles of the inclined circle we have defined [i.e. the ecliptic], in the opposite direction to the first motion.

If, then, we imagine a great circle drawn through the poles of both the above-mentioned circles, (which will necessarily bisect each of them, that is the equator and the circle inclined to it [the ecliptic], at right angles), we will have four points on the ecliptic: two will be produced by [the intersection of] the equator, diametrically opposite each other; these are called 'equinoctial' points. The one at which the motion [of the planets] is from south to north is called the 'spring' equinox, the other the 'autumnal'. Two [other points] will be produced by [the intersection of] the circle drawn through both poles; these too, obviously, will be diametrically opposite each other; they are called 'tropical' [or 'solsticial'] points. The one south of the equator is called the 'winter' [solstice], the one north, the 'summer' [solstice].

We can imagine the first primary motion, which encompasses all the other motions, as described and as it were defined by the great circle drawn through both poles [of equator and ecliptic] revolving, and carrying everything else with it, from east to west about the poles of the equator. These poles are fixed, so to speak, on the 'meridian' circle, which differs from the aforementioned [great] circle in the single respect that it is not drawn through the poles of the ecliptic too at all positions of the latter. Moreover, it is called 'meridian' because it is considered to be always orthogonal to the horizon. For a circle in such a position divides both hemispheres, that above the earth and that below it, into two equal parts, and defines the midpoint of both day and night.

The second, multiple-part motion is encompassed by the first and encompasses the spheres of all the planets. As we said, it is carried around by the aforementioned [first motion], but itself goes in the opposite direction about the poles of the ecliptic, which are also fixed on the circle which produces the first motion, namely the circle through both poles [of ecliptic and equator]. Naturally they [the poles of the ecliptic] are carried around with it [the circle through both poles], and, throughout the period of the second motion in the opposite direction, they always keep the great circle of the ecliptic, which is described by that [second] motion, in the same position with respect to the equator.

# Chapter Five

## *Philosophical Interlude: The "Save the Phenomena" Position*

Note: I am indebted to Professor André Goddu for his suggested revisions in the first section of this chapter and especially for having written its second section, which discuses early Lutheran responses to Copernicanism. MJC

### *On the History of the "Save the Phenomena" Position to Osiander*

Throughout much of history, one of the major positions in the methodology of science has been what is sometimes referred to as the "save the phenomena" position. In current parlance, this position is usually labeled "instrumentalism" or "fictionalism," although these modern terms are insufficiently capacious for our purposes. The basic tenet of this position is that judgments of the value of a scientific theory should be made in terms of only two criteria: the ability of the theory to save (account for or predict) the phenomena and the theory's simplicity. Among the criteria viewed as inappropriate is the realist's claim that a theory must be true. For an advocate of the save the phenomena position, a theory is nothing more than a useful fiction or instrument to account for phenomena. Although it is often assumed that the instrumentalist position is of recent origin, the following materials, which are based largely on Pierre Duhem's short book *To Save the Phenomena*,[1] demonstrate that as a strategy, although not as an exclusive philosophical view, instrumentalism was well known in antiquity. This position is of present relevance because a substantial portion of the Copernican debate concerned whether or not this methodology should be viewed as appropriate for astronomy.

At the present time, astronomy is distinguished from physics mainly on the basis of the different subject matters of these

[1] Pierre Duhem, *To Save the Phenomena: An Essay on the Idea of Physical Theory from Plato to Galileo*, trans. Edmund Doland and Chaninah Maschler (Chicago, 1969).

disciplines. In late antiquity, however, it became common to distinguish these areas in good part by reference to their supposedly different methodologies. This distinction is evident in the following passage from the first century A.D. author **Geminus,** who, probably under the influence of Aristotle's ideas, asserted:

> It is the business of physical inquiry to consider the substance of the heaven and the stars, their force and quality, their coming into being and their destruction, nay, it is in a position even to prove the facts about their size, shape, and arrangement; astronomy, on the other hand, does not attempt to speak of anything of this kind, but proves the arrangement of the heavenly bodies by considerations based on the view that the heaven is a real κόσμος, and further, it tells us of the shapes and sizes and distances of the earth, sun, and moon, and of eclipses and conjunctions of the stars, as well as of the quality and extent of their movements. Accordingly, as it is connected with the investigation of quantity, size, and quality of form or shape, it naturally stood in need, in this way, of arithmetic and geometry. The things, then, of which alone astronomy claims to give an account it is able to establish by means of arithmetic and geometry. Now in many cases the astronomer and the physicist will propose to prove the same point, e. g., that the sun is of great size or that the earth is spherical, but they will not proceed by the same road. The physicist will prove each fact by considerations of essence or substance, of force, of its being better that things should be as they are, or of coming into being and change; the astronomer will prove them by the properties of figures or magnitudes, or by the amount of movement and the time that is appropriate to it. Again, the physicist will in many cases reach the cause by looking to creative force; but the astronomer, when he proves facts from external conditions, is not qualified to judge of the cause, as when, for instance, he declares the earth or the stars to be spherical; sometimes he does not even desire to ascertain the cause, as when he discourses about an eclipse; at other times he invents by way of hypothesis, and states certain expedients by the assumption of which the phenomena will be saved. For example, why do the sun, the moon, and the planets appear to move irregularly? We may answer that, if we assume that their orbits are eccentric circles or that the stars

> describe an epicycle, their apparent irregularity will be saved; and it will be necessary to go further and examine in how many different ways it is possible for these phenomena to be brought about, so that we may bring our theory concerning the planets into agreement with that explanation of the causes which follows an admissible method. Hence we actually find a certain person, Heraclides of Pontus, coming forward and saying that, even on the assumption that the earth moves in a certain way, while the sun is in a certain way at rest, the apparent irregularity with reference to the sun can be saved. For it is no part of the business of an astronomer to know what is by nature suited to a position of rest, and what sort of bodies are apt to move, but he introduces hypotheses under which some bodies remain fixed, while others move, and then considers to which hypotheses the phenomena actually observed in the heaven will correspond. But he must go to the physicist for his first principles, namely, that the movements of the stars are simple, uniform, and ordered, and by means of these principles he will then prove that the rhythmic motion of all alike is in circles, some being turned in parallel circles, others in oblique circles.[2]

This passage shows that, at least in later antiquity, a sharp distinction was drawn between the methods of physics and astronomy. The task of astronomy was seen as being to account for or to save the celestial appearances by means of various mathematical devices. Put positively, this hardly seems radical, yet if examined more carefully, it becomes clear that this claim has far reaching implications. The method of astronomy, so-called because it developed chiefly, although not solely, within the astronomical tradition, is what has been called the "save the phenomena" position. A satisfactory theory, as defined by this position, need only "save the phenomena." It need not be true for we can have no certainty of the truth of an astronomical theory because another theory could be devised that would also save the phenomena. Preference should be given among competing theories to that theory found to be most accurate and most simple. At various times, it was suggested that such an approach is the

---

[2]As quoted in Thomas Heath, *Aristarchus of Samos: The Ancient Copernicus* (Oxford, 1913; New York: Dover, 1981), pp. 275–6.

only one possible for astronomy, which some viewed as treating the motions of divine entities.

Many of these features are evident in the Geminus quotation; they also appear in the writings of **Ptolemy,** who in his *Almagest* stated:

> If every apparent movement gets saved, as warranted by the hypotheses, why should anyone find it surprising that it is from such complicated motions that the movements of the heavenly bodies result?
>
> Let no one judge the real difficulties of the hypotheses in terms of the constructions we have devised. It is not fitting to compare things human with things divine. We should not base our trust in things so high on examples drawn from what is most greatly removed from them: For is there anything that differs more from changeless beings than beings that are constantly changing?[3]

This passage and others suggest that Ptolemy did not attribute reality to his eccentrics, epicycles, deferents, and equants; rather he viewed them as only mathematical devices to account for the phenomena.

The fifth century Neoplatonist **Proclus** also discussed this point; at the conclusion of his *Outline of the Astronomical Hypotheses*, Proclus warned:

> The astronomers who are eager to prove the uniformity of celestial motions are in danger of unconsciously proving that their nature is irregular and full of changes. What shall we say of the eccentrics and the epicycles of which they continually talk? Are these only inventions or have they a real existence in the spheres to which they are fixed? If they are only inventions, their authors have, all unaware, deviated from physical bodies into mathematical concepts and have derived the causes of physical motions from things that do not exist in nature. . . . But if the circles really exist, the astronomers destroy their connection with the spheres to which the circles belong. For they attribute separate motions to the circles and to the spheres and, moreover, motions that, as regards the

---

[3]As quoted in Duhem, *To Save the Phenomena*, p. 17.

> circles, are not at all equal but in the opposite direction. They confound their mutual distances and sometimes let them coincide in one plane, sometimes separate them and let them cross each other. Thus there will result all sorts of divisions, foldings-up and separations of the celestial bodies.
>
> Further, the account given of these mechanical hypotheses seems to be haphazard. Why, in each hypothesis, is the eccentric in this particular state—either fixed or mobile—and the epicycle in that, and the planet moving either in a retrograde or in a direct sense? What are the reasons for those planes and their separations—I mean the *real* reasons that, once understood, will relieve the mind of all its anguish?—this they never tell us. In fact, they proceed in a reverse order: they do not draw conclusions from their hypotheses like the other sciences, but instead they attempt to construct hypotheses which fit those conclusions which should follow from them. . . . However, one should bear in mind that these hypotheses are the most simple and most fitting ones for the divine bodies. They were invented in order to discover the mode of the planetary motions which in reality are as they appear to us, and in order to make the measure inherent in these motions apprehendable.[4]

Turning from the ancients to the medieval Arabs, we find that all Arabs were astronomical realists; nevertheless, they recognized, although they did not accept, the "save the phenomena" point of view. Consider **Averroes,** who discussed and criticized this position in his commentaries on Aristotle's *On the Heavens* and *Metaphysics*. In the former commentary, Averroes asserted:

> We find nothing in the mathematical sciences that would lead us to believe that eccentrics and epicycles exist.
>
> For astronomers propose the existence of these orbits as if they were principles and then deduce conclusions from them which are exactly what the senses can ascertain. In no way do they demonstrate by such results that the assumptions they have employed as principles are, conversely, necessities.[5]

[4]As quoted in S. Sambursky, *The Physical World of Late Antiquity* (London, 1962), pp. 148–9.

[5]As quoted in Duhem, *To Save the Phenomena,* p. 30.

In the latter commentary, Averroes stated:

> The astronomer must, therefore, construct an astronomical system such that the celestial motions are yielded by it and that nothing that is from the standpoint of physics impossible is implied. . . . Ptolemy was unable to see astronomy on its true foundations. . . . The epicycle and the eccentric are impossible. We must, therefore, apply ourselves to a new investigation concerning that genuine astronomy whose foundations are principles of physics. . . . Actually, in our time astronomy is nonexistent; what we have is something that fits calculation but does not agree with what is.[6]

Although the Arabs disapproved of this position, the twelfth-century Jewish philosopher **Moses Maimonides** in his *Guide of the Perplexed* was more favorably inclined. In the following passage Maimonides gave an excellent statement of this position:

> Know with regard to the astronomical matters mentioned that if an exclusively mathematical-minded man reads and understands them, he will think that they form a cogent demonstration that the form and number of the spheres is as stated. Now things are not like this, and this is not what is sought in the science of astronomy. Some of these matters are indeed founded on the demonstration that they are that way. Thus it has been demonstrated that the path of the sun is inclined against the equator. About this there is no doubt. But there has been no demonstration whether the sun has an eccentric sphere or an epicycle. Now the master of astronomy does not mind this, for the object of that science is to suppose as a hypothesis an arrangement that renders it possible for the motion of the star to be uniform and circular with no acceleration or deceleration or change in it and to have the inferences necessarily following from the assumption of that motion agree with what is observed. At the same time the astronomer seeks, as much as possible, to diminish motions and the number of the spheres. For if we assume, for instance, that we suppose as a hypothesis an arrangement by means of which the observations regarding the motions of one particular star can be accounted for through the assumption of three

[6]As quoted in Duhem, *To Save the Phenomena,* p. 31.

> spheres, and another arrangement by means of which the same observations are accounted for through the assumption of four spheres, it is preferable for us to rely on the arrangement postulating the lesser number of motions. For this reason we have chosen in the case of the sun the hypothesis of eccentricity, as Ptolemy mentions, rather than that of an epicycle.[7]

In the next passage, Maimonides suggested additional reasons for adopting this position as well as ways in which it could be reconciled with religion.

> However, regarding all that is in the heavens, man grasps nothing but a small measure of what is mathematical; and you know what is in it. I shall accordingly say in the manner of poetical preciousness: *The heavens are the heavens of the Lord, but the earth hath He given to the sons of man.* [Psalms 114:16] I mean thereby that the deity alone fully knows the true reality, the nature, the substance, the form, the motions, and the causes of the heavens. But he has enabled man to have knowledge of what is beneath the heavens, for that is his world and his dwelling-place in which he has been placed and of which he himself is a part. This is the truth. For it is impossible for us to accede to the points starting from which conclusions may be drawn about the heavens; for the latter are too far away from us and too high in place and in rank.... And to fatigue the minds with notions that cannot be grasped by them and for the grasp of which they have no instrument, is a defect in one's inborn disposition or some sort of temptation.[8]

The medieval philosophers of the Latin West recognized this position, although most did not accept it. **Thomas Aquinas**, for example, in his *Summa Theologica*, Q. 32, Pt. I., art. I, reply to objection 2, stated:

> Reasoning may be brought forward for anything in a two-fold way: firstly, for the purpose of furnishing sufficient proof of some principle, as in natural science, where sufficient proof

[7]As quoted in Duhem, *To Save the Phenomena,* p. 34.
[8]As quoted in Duhem, *To Save the Phenomena,* p. 35.

> can be brought to show that the movement of the heavens is always of a uniform velocity. Reasoning is employed in another way, not as furnishing a sufficient proof of a principle, but as showing how the remaining effects are in harmony with an already posited principle; as in astronomy the theory of eccentrics and epicycles is considered as established, because thereby the sensible appearances of the heavenly movements can be explained; not, however, as if this proof were sufficient, since some other theory might explain them.[9]

Among the other medieval philosophers, **Roger Bacon** leaned toward an Aristotelian realism, whereas **Bernard of Verdun** preferred a Ptolemaic realism. **Peter of Padua**, on the other hand, adopted a position essentially identical to that advocated by Ptolemy in his *Almagest*.

An important modification and extension of the "save the phenomena" position was advocated by **Nicholas of Cusa** in the fifteenth century and by **Luiz Coronel** in the early sixteenth century. These two authors adopted the "save the phenomena" position, not only for astronomy, but for physics as well.

In summarizing and commenting on these ancient and medieval views, let us return to our earlier distinction between an instrumentalist strategy and an instrumentalist philosophy of science. There seems little reason to doubt Pierre Duhem's conclusion that certain mathematical hypotheses in the Middle Ages were treated as mere devices—truth was not attributed to them. On the other hand, there is little evidence to support the further conclusion that natural philosophers who treated such mathematical devices formally were committing themselves thereby to an instrumentalist philosophy of science. The Averroistic rejection of hypotheses and evidently of the "save the phenomena" principle logically followed from acceptance of the reality of Aristotelian physics and cosmology. On the other hand, those who rejected Aristotle's model interpreted the Ptolemaic as true. Similarly, Thomas Aquinas, who was surely a realist, regarded the astronomical hypotheses of the Ptolemaic system as

---

[9]As given in *Basic Writings of Thomas Aquinas*, ed. and annotated by Anton C. Pegis, vol. I (New York, 1945), p. 318.

not necessarily true, yet he had no doubt about the truth of the principles of Aristotelian physics and cosmology.

In any case, the medieval interpretation of mathematical hypotheses as mere devices was at best instrumentalist as a strategy or response to the demanding standards of a **realist** epistemology bound to acceptance of the Aristotelian cosmos. The question of the truth of hypotheses was settled by reference to perceivable phenomena. In the case of astronomy the problem of observational accuracy alone is sufficient to understand why so many medieval philosophers were reluctant to conclude for the absolute truth of mathematical devices. Such reluctance is a testimony to their fundamental commitment to realism. Medieval natural philosophers accepted the predictive adequacy of theoretical statements not because they were uninterested in truth and certainly not because they believed that truth as a correspondence with reality is unattainable, but because they regarded hypotheses as at best contingent and because they were committed to inductive canons of justification. Consequently, medieval natural philosophers resigned themselves in the face of invariable speculative principles to the limitations that they had placed on achieving objective knowledge, limitations that they believed were inherent and, therefore, insurmountable in the relation between the knower and the known.

When *De revolutionibus orbium coelestium*, the great astronomical treatise of **Nicholas Copernicus,** was first published in 1543, it contained an unsigned preface written by **Andreas Osiander**. In that preface, Osiander adopted the save the phenomena position as a basis for urging readers to give careful consideration to the shocking ideas of Copernicus. Without the materials discussed up to this point, Osiander's preface might seem incomprehensible. Given these materials, however, we can see Osiander's statement as essentially a reassertion of one traditional position regarding astronomical method. Osiander's approach was not that of Copernicus, who viewed his system in realist terms. Galileo and Kepler were also realists; in fact, Kepler stressed that Osiander, not Copernicus, had penned this preface. Nonetheless, Osiander's position was favored by some participants in the Copernican controversy, for example, **Cardinal Robert Bellarmine,** who used it against Galileo. The next reading consists of Osiander's famous preface:

## To the Reader
## Concerning the Hypotheses of This Work

There have already been widespread reports about the novel hypotheses of this work, which declares that the earth moves whereas the sun is at rest in the center of the universe. Hence certain scholars, I have no doubt, are deeply offended and believe that the liberal arts, which were established long ago on a sound basis, should not be thrown into confusion. But if these men are willing to examine the matter closely, they will find that the author of this work has done nothing blameworthy. For it is the duty of an astronomer to compose the history of the celestial motions through careful and expert study. Then he must conceive and devise the causes of these motions or hypotheses about them. Since he cannot in any way attain to the true causes, he will adopt whatever suppositions enable the motions to be computed correctly from the principles of geometry for the future as well as for the past. The present author has performed both of these duties excellently. For these hypotheses need not be true nor even probable. On the contrary, if they provide a calculus consistent with the observations, that alone is enough. Perhaps there is someone who is so ignorant of geometry and optics that he regards the epicycle of Venus as probable, or thinks that it is the reason why Venus sometimes precedes and sometimes follows the sun by forty degrees and even more. Is there anyone who is not aware that from this assumption it necessarily follows that the diameter of the planet at perigee should appear more than four times, and the body of the planet more than sixteen times, as great as at apogee? Yet this variation is refuted by the experience of every age. In this science there are some other no less important absurdities, which need not be set forth at the moment. For this art, it is quite clear, is completely and absolutely ignorant of the causes of the apparent nonuniform motions. And if any causes are devised by the imagination, as indeed very many are, they are not put forward to convince anyone that they are true, but merely to provide a reliable basis for computation. However, since different hypotheses are sometimes offered for one and the same motion (for example, eccentricity and an epicycle for the sun's motion), the astronomer will take as his first choice that hypothesis which is the easiest to grasp. The philosopher will perhaps seek the semblance of the truth. But neither of them will understand or state anything certain, unless it has been divinely revealed to him.

Therefore alongside the ancient hypotheses, which are no more probable, let us permit these new hypotheses also to become known, especially since they are admirable as well as simple and bring with them a huge treasure of very skillful observations. So far as hypotheses are concerned, let no one expect anything certain from astronomy, which cannot furnish it, lest he accept as the truth ideas conceived for another purpose, and depart from this study a greater fool than when he entered it. Farewell.[10]

*Lutheran Responses to Copernicus*

Andreas Osiander's tactic has been the subject of much controversy. In analyzing Osiander's preface, one should be aware not only of its apparent instrumentalism, but also of the polemically religious atmosphere of the sixteenth century. When set in that context, Osiander's "Letter" may appear to be somewhat less pernicious in its intent and effects.

After all, the Copernican hypothesis challenged Aristotelian physics and the literal interpretation of several passages of Scripture. It was, of course, on these grounds that theologians, Catholic and Protestant, rejected the theory in its literal form. Although there are doubts about the authenticity of some of the remarks attributed to **Martin Luther** in his *Table Talk (Tischreden)*, the version that has been cited has Luther saying the following: "The fool wants to turn the whole art of astronomy upside down." Even if unhistorical, the account is not implausible as the kind of reaction one might expect from a reformer who set so much store in the authority of Scripture.

Still, Luther's alleged remark came at a time (1539) when only a sketch of the Copernican theory was available. Luther seems not

[10]As given in Nicholas Copernicus, *On the Revolutions*, ed. by Jerzy Dobrzycki, trans. with commentary by Edward Rosen (Baltimore, 1978), p. XVI.

to have reacted to the complete account of the theory published two years before Luther's death. Luther's education at the University of Erfurt does not help us in settling the question of his genuine view. As a student of arts he was exposed to philosophy and astronomy, but of the arts of the quadrivium it was music above all, not astronomy, that claimed his interest and extracurricular activities. His philosophical education was clearly Aristotelian in its essentials, although he was very likely instructed in an interpretation of it that emphasized the contingency of the Aristotelian cosmos. Still, this would have convinced Luther at best that astronomical theory was beyond certain, empirical verification.

Among the circle of **Philip Melanchthon**, one of Luther's followers at the University of Wittenberg, a response to the Copernican theory was developed that permitted the theory to be used without acceptance of its claim that the sun is literally at the center of the universe. This version of the theory has become known as the Wittenberg Interpretation, and its role in the Copernican revolution has itself become a matter of controversy.

At the center of this development stands the figure of Philip Melanchthon. Melanchthon was disposed early in life toward humanistic and Greek studies, and he developed a deep interest in astrology and astronomy. An active leader in the humanistic movement at the University of Tübingen, he assumed a professorship of Greek at the newly formed University of Wittenberg in 1518. Some of Luther's more zealous followers at Wittenberg demanded not only a return to the simplicity of the early Church but the abolition of all education.

Against these trends Melanchthon undertook a campaign of educational reform, which influenced nearly every university in Germany. Next to the humanistic disciplines, Melanchthon's textbooks in physics and astronomy gave mathematics a special place in the curriculum. Wittenberg seems to have supplied all of the other German universities with astronomers and mathematicians. Melanchthon's own view of the Copernican theory was initially harsh and critical, but eventually his typical capacity for flexibility and compromise brought him to see pragmatic advantages in the theory, even if it were false because in conflict with Scripture. As a matter of fact, this reading of the

Copernican system derived from Osiander's "Letter." As we have seen, on this view the models were to be viewed as merely convenient devices for making astronomical calculations. Melanchthon's partial acceptance of this moderate position permitted the Copernican system space and time to be developed, but it also masked its truly revolutionary implications.

The two most important members of the Melanchthon circle were **Erasmus Reinhold** and **Caspar Peucer**. Reinhold shared Copernicus's dissatisfaction with the Ptolemaic theory, but his adoption of the mathematics was accompanied by a perfect neutrality concerning the problem of geocentrism and heliocentrism. Through Reinhold the image of Copernicus the calculator and reformer of equant-ridden astronomy was etched indelibly on the minds of Wittenberg astronomers. Peucer succeeded in transmitting this Wittenberg interpretation to the other universities in Germany. As with Reinhold his teacher, Peucer ignored the cosmological implications of the Copernican theory. At least up to 1570 this reading of Copernicanism prevailed among the German universities.

Our concern, of course, involves the controversy surrounding the Wittenberg Interpretation with respect to the following questions:

(1) Did this interpretation constitute an instrumentalist reading of a scientific theory?

(2) Why did the next generation of astronomers react against their predecessors and emphasize the cosmological consequences of the Copernican system?

(3) How could Copernicanism in this latter version be upheld without contradicting Scripture?

As we saw, the Melanchthon circle rejected the literal heliocentrism of the Copernican system. Its members saw the advantages of the system for purposes of calculation and because it eliminated the equant. So far this sounds instrumentalist. But what has been overlooked here are the reasons for the rejection of heliocentrism—the fact that it contradicts Scripture and that there was no empirical evidence that supported the theory. In short, one of the standards whereby Copernicanism was rejected as literally true was a clearly realist criterion, namely, the presence or absence of empirical verification. Without such absolutely certain empirical verification, it was extremely unlikely that anyone would lightly accept the consequence that Scripture would have to be reinterpreted.

The answer to the first question, then, is no; the Wittenberg Interpretation was not instrumentalist. It rejected the Copernican theory as false and, it is true, used it for certain advantages in calculation, but the Melanchthon circle continued to maintain the literal truth of the geocentric hypothesis. The result was that they downplayed the revolutionary cosmology of the system and thereby permitted it breathing room, allowing for acceptance of the theory for limited advantages in calculation.

It seems as well that this acceptance of the Copernican system as a reform, not a revolution, helps to put in perspective the reactions of the next generation beginning roughly around 1570. Those reactions stemmed from two significant developments—(1) evidence that, although it did not confirm the Copernican theory, created even more problems for Ptolemaic astronomers, and (2) the willingness on the part of a few astronomers to accept the Copernican system as literally true not on the basis of any new empirical evidence but on the basis of a partly resurrected view of the universe that combined mathematical reasoning with Neoplatonic conceptions of the mathematical harmony of the universe, conceptions that also gave an increased importance to the role of the sun in the cosmos. The new evidence was the appearance of comets in the 1570s, which threw many, until then, confirmed Ptolemaic astronomers into a dither. This did not in itself motivate them to accept Copernicanism because, as a matter of fact, the orbits of the comets were also a problem for Copernicans, but these new observations stimulated more movement away from the Ptolemaic theory. Still, the truly significant breakthrough in the 1570s was the fact that more astronomers were coming to grips with the cosmological implications of the Copernican theory. In short, these astronomers were at last willing to go beyond reformation and push for revolution. Not incidentally, Kepler was born in 1571 and Galileo just seven years earlier in 1564.

The fact that—beginning with the generation that centers around 1570—there were more astronomers willing to accept Copernicanism literally on the basis of its alleged mathematical simplicity and metaphysical coherence meant necessarily the serious adoption of a new approach to the biblical text. Actually, the principles were extremely old, especially the admonition that the Bible speaks in the language of men, that is to say, the principle of accommodation that had been used since time immemorial to explain anthropomorphisms, for example. The serious question was rather this: On what evidence should one reject the literal interpretation of the biblical text in favor of some other interpretation? The answer is the very same one that separated the reformers from the revolutionaries.

This is why we can make sense out of the reception of the Copernican theory only if we set it in the context of two competing, but fundamentally realist views: the empiricism of the Melanchthon circle versus the mathematical rationalism of later generations. For the former, as also for **Cardinal Bellarmine** in his dispute with Galileo nearly sixty years later, only indubitable demonstration of the truth of the Copernican theory could justify reinterpretation of the clear meaning of Scripture.

If the Melanchthon circle and Cardinal Bellarmine were instrumentalists, they would hardly have set indubitable demonstration as their criterion for their acceptance of the theory as true. Bellarmine's view appears very clearly in a letter dated 9 April 1615 to Paolo Antonio Foscarini. Bellarmine demanded a true demonstration, which no one had yet produced. As for the argument based on the relativity of appearances, Bellarmine responded:

> Anyone who departs from the beach, though to him it appears that the beach moves away, yet knows that this is an error and corrects it, seeing clearly that the ship moves and not the beach; but as to the sun and earth, no sage has needed to correct the error, since he clearly experiences that the earth stands still and that his eye is not deceived when it judges the sun to move, just as he is likewise not deceived when it judges that the moon and the stars move. And that is enough for the present.[11]

[11]As quoted in Stillman Drake, *Discoveries and Opinions of Galileo* (New York, 1957), p. 164.

By "indubitable demonstration" was meant Aristotelian demonstration, and every such demonstration requires a premise that is an absolutely certain principle derived from empirical evidence or that is self-evident. Bellarmine's objections show clearly that he expected empirical evidence as proof. Again, it seems that the instrumentalism of Melanchthon, Osiander, and Cardinal Bellarmine was a response conditioned by indubitable and clearly realist conclusions derived from Aristotle and Scripture.

## *Conclusion*

This discussion of the nature of the so-called "save the phenomena" or "instrumentalist" or "fictionalist" position in methodology has also documented the antiquity and attractiveness of that position and sketched its history from antiquity to the early seventeenth century. It has also raised a number of important questions:

(1) What are the chief canons of this methodological position?

(2) What are the main arguments for and against it?

(3) Has the position proved beneficial in the development of science? If so, in precisely what ways? If not, in what ways?

(4) If a theory's simplicity and its efficacy in saving phenomena are not adequate grounds for its acceptance, what are?

# Chapter Six

## *The Copernican System*

**Nicholas Copernicus**

### *Introduction*

A number of remarkable features make the Copernican revolution an especially interesting subject of study. That revolution was launched in 1543 by a canon of Frauenburg Cathedral, located in a remote corner of Europe. Its manifesto was a book so teeming with technicalities that only the most sophisticated attempted to read it and only the most daring of these espoused its message. Although medieval scholars had made numerous contributions to astronomy, the materials with which Nicholas Copernicus (1473–1543) constructed his new system had been available almost without exception for far more than a millennium, being essentially the same elements employed by Ptolemy in erecting the edifice for which he was renowned. In fact, as this chapter reveals, Copernicus, that remarkable revolutionary, was in many ways conservative to the core of his being. And in one sense, the new theory he presented differed significantly in only a single mathematical detail from the Ptolemaic system: Copernicus, as it were, proposed that a new point be taken as the center of the coordinate system used by astronomers. Nonetheless, the change from a geocentric and geostatic system to a heliocentric system had profound implications that extended far beyond astronomy. It transformed the earth into a planet, humans into planetarians, and stars into suns, thereby "infinitizing" the universe. And it raised deep and disturbing philosophical and theological issues. The poet Goethe, writing nearly three centuries later, summarized the significance of the revolution wrought by this conservative Polish canon:

> Humanity has perhaps never faced a greater challenge; for by his admission [that humanity is not the center of the universe], how much else did not collapse in dust and smoke: a second paradise, a world of innocence, poetry and piety, the witness of the senses, the conviction of a religious and poetic faith. . . .[1]

## *Life of Copernicus*

Nicholas Copernicus was born on February 19, 1473, in Toruń, a town under the Polish king. He was the youngest of the four children in a well-to-do merchant family. After his father's death in 1483, Copernicus came under the care of Lucas Watzenrode, his maternal uncle, who in 1489 became Bishop of Warmia. In 1491, Copernicus entered Cracow University, where

[1] J. W. Goethe, *Zur Farbenlehre* in Goethe, *Sämtliche Werke*, vol. 40 (Stuttgart, 1902), p. 185.

he was introduced to the sciences. Shortly after this, he became a canon of the cathedral at Frauenburg in Poland. Although remaining a canon, he never chose to be ordained to the priesthood. In 1496, he proceeded to Italy, entering the University of Bologna, where he studied law and increased his knowledge of astronomy by close contacts with the astronomer Domenico Maria di Novara. In 1501, after returning for a short period to Poland, he resumed his education by studying medicine at Padua University. He was awarded a doctorate in canon law in 1503 by another Italian university, the University of Ferrara. He then returned permanently to Poland, living from 1504 to 1510 in Heilsberg with Bishop Watzenrode, whence he proceeded to Frauenburg, where for the most part he remained until his death in 1543, performing various administrative duties for the chapter of canons at Frauenburg Cathedral, sometimes acting as a physician, and, largely in isolation, remodeling the universe.

Good evidence indicates that Copernicus created his heliocentric system in the period from 1510 to 1514, during which time he presented that system in a short manuscript, known as his *Commentariolus,* which, although not published during his lifetime, he circulated selectively among interested scholars. It was also during this period that he established a room for astronomical observation at the Cathedral and made the first astronomical observation recorded in the full scale exposition of his system that he finally published in 1543, his *De revolutionibus orbium coelestium.*

Copernicus composed this classic treatise around 1530, adding minor modifications during the remaining thirteen years of his life. Word of his new system gradually spread even before its publication. For example, in 1533, Johann Widmanstadt, the Papal Secretary, explained the Copernican system to Pope Clement VII (d. 1535) as well as to two cardinals who were present. Three years later, Cardinal Schönberg (d. 1537), to whom Widmanstadt had become secretary, wrote Copernicus from Rome, asking him to send an explanation of his new system. In 1539, Bishop Dantiscus of Warmia ordered three canons at Frauenburg, including Copernicus, to separate themselves from their mistresses. Later in that year, Georg Joachim Rheticus (1514–1574), a Lutheran astronomy professor, came to study with Copernicus. During the summer of 1539, Copernicus and Rheticus visited Copernicus's friend Bishop Tiedemann Giese (d. 1550), who urged Copernicus to publish his system.

In 1540, Rheticus published his *Narratio prima*, a short work that was the first published exposition of the Copernican system. Copernicus, having by this time decided to publish his book, began making final revisions, including the preparation in 1542 of its dedicatory letter to Pope Paul III. Arrangements were made for the publication of the work by a Nuremberg publisher, with Rheticus assuming responsibility for seeing the manuscript through the press. When the latter plan was disrupted by Rheticus in 1542 accepting a position at the University of Leipzig, a Lutheran theologian at Nuremberg, Andreas Osiander (1498–1552), took responsibility for the manuscript. Moreover, concerned about the reception of its radical central thesis, Osiander, without the approval of Copernicus, added an unsigned introductory letter in which he proposed that the system be viewed not necessarily as true, but rather as a hypothetical device that would "save" the celestial phenomena.

In December, 1542, Copernicus suffered a cerebral hemorrhage that resulted in partial paralysis. Solid but less than fully conclusive evidence supports the dramatic story that the first copies of *De revolutionibus* reached Copernicus on May 24, 1543, the day on which he died. If he, in fact, saw the work, he would have been disturbed by Osiander's unauthorized preface, which seriously misrepresented Copernicus's conviction that heliocentrism is the true theory of the world. Bishop Giese, aware of Copernicus's conviction, protested to the publisher, urging that a corrected edition be issued, with Osiander's letter removed. Giese's effort failed, which tended to perpetuate the erroneous belief that Copernicus viewed his system only as a mathematical device. Eventually, various Copernicans, most notably Kepler, called attention to Osiander's authorship.

### *The Origin of and Evidence for Copernicus's System*

Two fascinating questions concerning Copernicus's system are: (1) how did he come to create that system? and (2) what evidence convinced him of its correctness?

Because Copernicus left no clear record of the process by which he had arrived at his new theory, finding an answer to the first question is difficult. Although no universally accepted answer to it has emerged, various points are clear. Copernicus did not derive his system from new astronomical observations or instruments nor did he possess mathematical advances unavailable to Ptolemy. The materials presented up to this point should make clear that because of the problem of the relativity of motion, no observation, even one made with the telescope (which was invented more than a half century after the death of Copernicus) of the objects in our planetary system could demonstrate the correctness of the heliocentric theory. Moreover, Copernicus recorded in his book only twenty-seven observations that he had made. One of the ironies of the situation is that an understanding of the mathematics discussed up to this point reveals that mathematical considerations could not determine which system is correct. For this reason, it should be clear why some scholars have characterized the choice between the two systems as one that rested on aesthetic considerations. Moreover, Copernicus possessed no new mathematical techniques. His book is filled with the traditional mathematical devices of later Greek astronomy; eccentrics, deferents, and epicycles fill its pages. Copernicus claimed to have avoided use of the equant, but this has been challenged. It is also clear that Copernicus did not simply revive Aristarchus's theory; in fact, he never mentioned that ancient heliocentrist in the published version of his *De revolutionibus*. Various theories as to how Copernicus did, in fact, arrive at his system have been put forward, but a consideration of these proposals is unnecessary for present purposes.

Finding an answer to the second question, that concerning the evidence that supported Copernicus's confidence in his system, is also complicated by the problem of the relativity of motion. That problem accounts for the fact that the Copernican system achieved no greater predictive accuracy than the Ptolemaic system, at least in the improved versions of the Ptolemaic system available in the sixteenth century. Copernicus did claim that his system was simpler than Ptolemy's, even stating in his *Commentariolus* that his system required far fewer epicycles than those used by his ancient predecessor. Nonetheless, a careful comparison of the two systems shows that this claim is not justifiable. Copernicus used numerous epicycles, which he needed to account for the non-uniformity of the planets' motions as well as for their motions in latitude. Despite these obstacles, Copernicus was able to cite some evidences for his new theory. These emerge from a knowledge of his system and from the selection from Copernicus's book presented later in this chapter.

*The Copernican Theory of the Stars, Sun, and Moon*

In the Copernican system, the starry vault does not rotate; rather the **earth rotates**, making the stars appear to revolve around the earth. The earth's axis of rotation passes through its north and south poles. Copernicus claimed that the daily rotation of the starry vault is simply an apparent effect resulting from the earth's daily rotation. Problem 2 in Chapter One was devised to illustrate this point.

According to Copernicus, the earth **revolves around the sun** with a period of a year, whereas the sun remains fixed in position. Problem 6 in Chapter One, as well as the next diagram, illustrate that this produces the same appearances as the sun orbiting the earth.

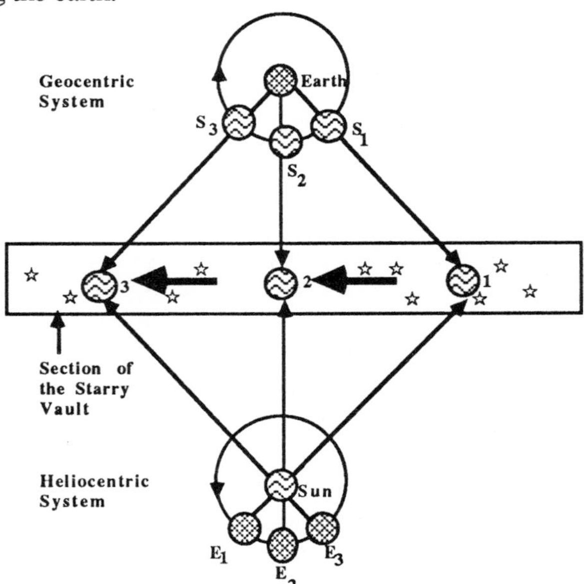

In both systems, the sun is seen to move from position 1 to positions 2 and 3 on the starry vault. Proponents of the Ptolemaic theory attributed this to the sun moving from position $S_1$ to $S_2$ and $S_3$, whereas Copernicans maintained that this appearance is due the earth moving from $E_1$ to $E_2$ and $E_3$. This also suggests that the Copernican explanation of the seasons is essentially identical to the geocentric explanation. Because his treatment of the precession of the equinoxes was complex yet not directly relevant to the choice between his system and Ptolemy's, it need not be discussed.

On one level, the Ptolemaic and Copernican theories of the **moon** are identical. For example, each has the moon revolve around the earth and the lunar phases are explained in an identical manner. It turns out, however, that Copernicus devised a superior mathematical model for the moon, which did not predict (as Ptolemy's had) a near doubling of the moon's apparent size. Their lunar theories need not, however, concern us because they have no direct bearing on the core elements of the two systems.

### *The Copernican Theory of the Planets*

Problems 12 and 13 of Chapter One reveal that at least on the first level the problem of accounting for the planetary motions is linked to the problem of the relativity of motion, the principle that the motion of any object can only be defined in reference to some other object that is assumed to be at rest. Imagine a basketball player constantly tracking the distances of nine other players. Every other player is similarly engaged in monitoring the motions of the remaining players as seen from his or her perspective. Does anyone on the court have a privileged position in this process of tracking positions and distances? Is any one player's description of these motions superior to others? Obviously not. This suggests that with sufficiently complex mathematical devices, the celestial motions can be accurately and precisely described from anywhere. Copernicus's bold leap was to describe the motions from the point of view of the sun, which he assumed to be stationary.

Copernicus claimed that each planet, including the earth, revolves around the sun. Moreover, he asserted that the **ordering**

**of the planets** is Mercury, Venus, earth, Mars, Jupiter, and Saturn, i.e., Mercury has the smallest and Saturn the largest orbit with the orbit of the earth being between those of Venus and Mars. The larger the orbit, the longer the period assigned to it.

*The Copernican Explanation of Bounded Elongation*

As previously noted, Mercury and Venus have bounded elongation; that is, their zodiacal positions are always within 28° and 46° respectively of the sun. Ptolemy **accounted** for this by centering the epicycles of Mercury and Venus on a line through the sun. Copernicus claimed not only to account for this phenomenon, but also to **explain** it. In other words, whereas Ptolemy's model could predict that Mercury and Venus remain near the sun, it could offer no reason why their motions should be linked to the sun in this curious manner. Copernicus not only predicted this effect, but also explained why it occurs; in particular, he suggested that these two planets revolve around the sun in orbits inside the earth's orbit (see diagram).

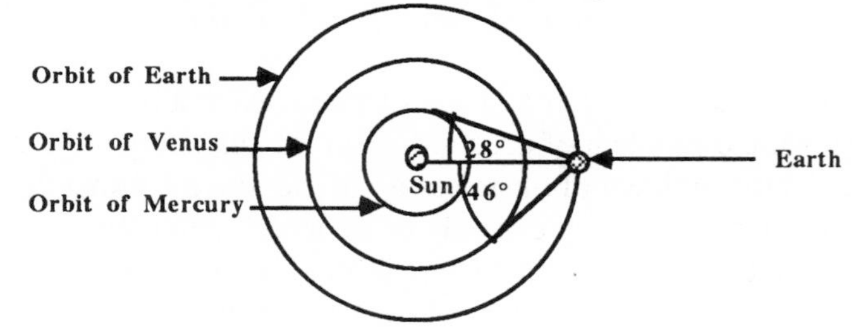


This methodological difference between the two systems raises the important philosophical question whether it should count toward the credibility of one theory as compared to another if the first theory, although unable to give greater predictive accuracy, can offer an explanation of phenomena.

*The Copernican Explanation of Retrograde Motion*

According to the Copernican system, the planets never actually retrogress, that is, they never back up in their orbital motions. Nonetheless, as every planetary observer knows, the planets do appear to retrogress. How is this to be explained? The next diagram, although somewhat simplified as compared to the actual situation, explains this. Let the earth (E) move around the sun. Let P in the diagram be an outer planet. As such, it will have a longer period of revolution than that of the earth. Let $E_1$, $E_2$,..., $E_{11}$ represent eleven successive positions of the earth and let $P_1$, $P_2$,..., $P_{11}$ represent eleven successive positions of P, the outer planet. The starry vault is represented at the right. To see where against the background of the fixed stars P will appear as seen from E, draw lines from the corresponding positions of E and P, extending these lines to the starry vault. Mark these positions $V_1$, $V_2$,..., $V_{11}$. The line $E_6P_6V_6$ has already been drawn. Then note the order in which the points $V_1$, $V_2$,..., $V_{11}$ stand in relation to each other.

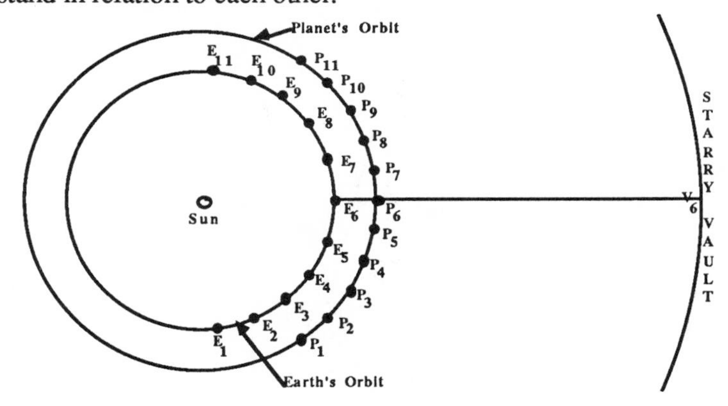


As this diagram indicates, the Copernican theory entails the claim that Mars, Jupiter, and Saturn will retrogress only when they are in opposition with the sun. This point is also illustrated in the next diagram, which consists of a comparison of how the retrogressions of Mars, Jupiter, and Saturn are generated in the two systems. In the Ptolemaic system, the retrogression of the planet P is produced by the rotations of the large epicycle on which the planet is positioned. The Copernican conception of the retrogression is that it is actually an appearance produced by the motion of the earth around the sun. Note that the size of the Ptolemaic epicycle in the upper part of the diagram is identical to that of the earth's orbit in the lower part of the figure, which further suggests that the Ptolemaic epicycles are in a sense reflections of the earth's orbital motion.

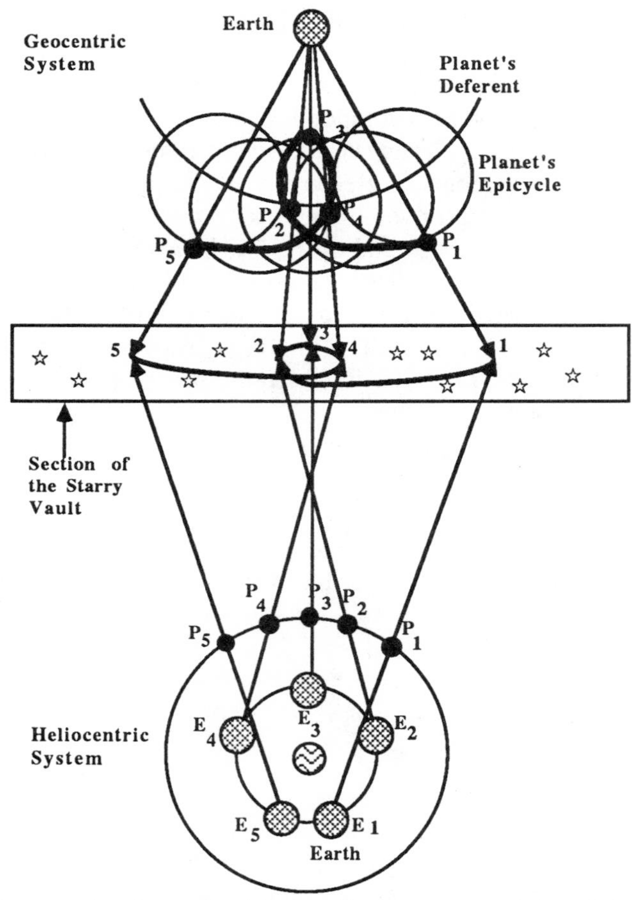


Another point to be noticed is that Copernicus could explain why these three planets retrogress only when in opposition with the sun. The status of Ptolemy's theory was significantly different. Like Copernicus, he was able to predict that Mars, Jupiter, and Saturn will retrogress when at opposition. This requirement was built into his system by his stipulation that the radius of the main epicycle for each of these three planets must remain parallel to a line from the earth to the sun. Nonetheless, because Ptolemy could give no reason why their motions should be linked in this curious manner to the motion of the sun, he could not explain why this should be the case.

*Distances of the Planets in the Copernican System*

Ptolemy had no way to calculate the actual or relative distances of the planets. In Copernicus's system, however, good estimates of **relative** distances are obtainable, but it is important to realize that these distances can be assumed correct **only if** the Copernican theory is true. If a person sees a winged object in the sky and assumes that it is a bird, the person can on this assumption estimate the object's distance. If, however, the winged object is an airplane, then the person's distance determination will be groundless. To illustrate the Copernican method for determining the distance of planets, let us calculate the radius of Venus's orbit, assuming that the radius of the earth's orbit is 10 units. Let Venus be at its greatest angular distance (46°) from the sun. This is represented in the next diagram.

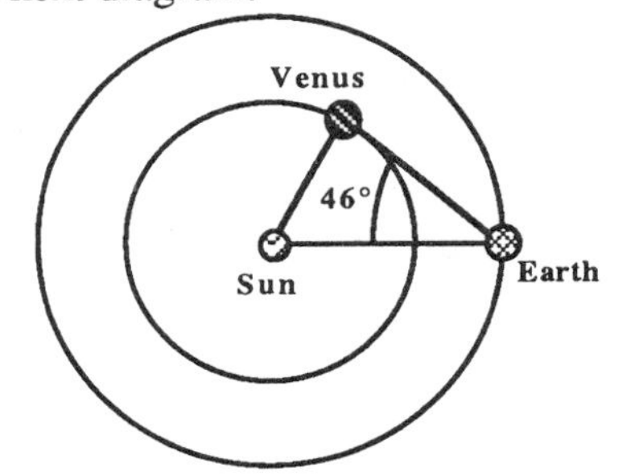


Line VE is tangent to Venus's orbit and hence perpendicular to VS, the radius of Venus's orbit. We know that $\angle SVE$ is 90° and that $\angle VES = 46°$ and are given that SE = 10. By trigonometry, VS/SE = VS/10 = sin 46°, but sin 46° = .72. Thus the radius of Venus's orbit is about 7.2 units.

The distance of an outer planet can also be calculated.

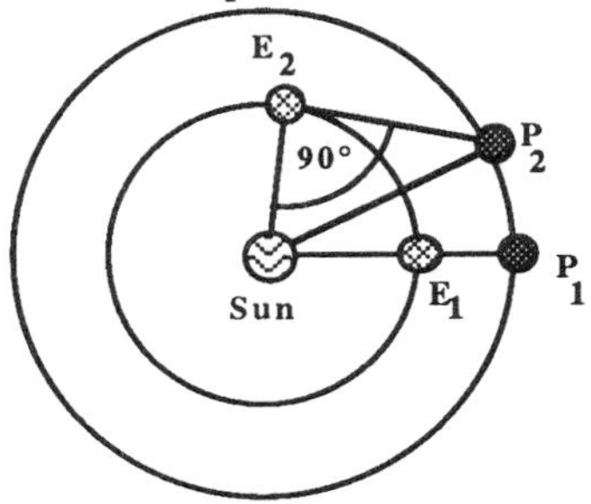

Let the planet first be in opposition with the earth. Then let the earth and planet move until the planet is 90° from the sun, that is, when $\angle P_2E_2S = 90°$. We thus have a right angled triangle and can apply trigonometry. We are able to calculate $\angle P_1SP_2$ from the formula: $\angle P_1SP_2/360° = (t_2 - t_1)/T_p$ where $t_2 - t_1$ is the time it takes the earth to move from $E_1$ to $E_2$ and also the time the planet takes in moving from $P_1$ to $P_2$. $T_p$ is the orbital period of the planet. We can also calculate $\angle E_1SE_2$; we have $\angle E_1SE_2/360° = (t_2 - t_1)/1$ year. Having obtained these values, we can determine $\angle E_2SP_2$: $\angle E_2SP_2 = \angle E_1SE_2 - \angle P_1SP_2$. Taking the radius of the earth's orbit as 10, we have for the radius of the planet's orbit R: $\cos\angle E_2SP_2 = SE/SP = 10/R$ or $R = 10/\cos\angle E_2SP_2$.

*Why Did the Ptolemaic System Work So Well?*

This question can be answered in a number of ways and on a variety of levels. Let us at present confine ourselves to asking why the Ptolemaic epicycle-deferent system worked so effectively, even though the true system of the world is the heliocentric system. The answer can be seen from problems 12 and 13 of Chapter One. Consider the case where it is given that three bodies A, B, and C exist and are located in such a manner that the position of C is fixed, whereas B and A move around C as indicated in the diagram at the left. Let us now consider the case where C is the sun, B is Venus, and A the earth.

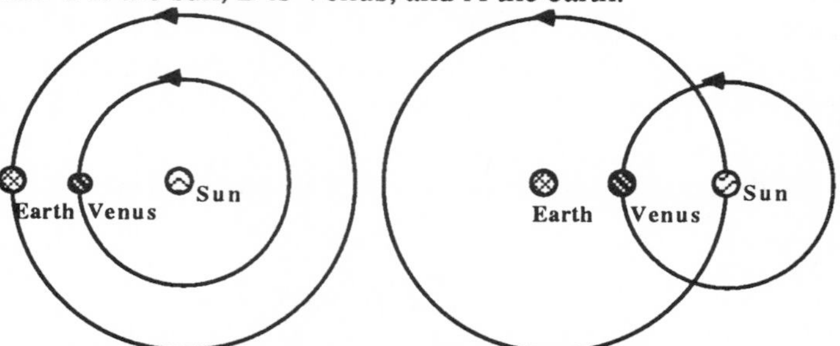


Viewed from earth, the sun always remains at the distance SE; moreover, S will have the same period of revolution around E as E would appear to have if seen from S. We can effectively represent the motion of Venus by having it move on an epicycle of radius SV, which epicycle (centered on S) moves on a deferent of radius SE (diagram on right). This arrangement is essentially Ptolemy's basic model for Venus. In effect, Ptolemy's main Venusian epicycle replicates the motion of Venus around the sun; of course, Ptolemy had no way of knowing that the Venusian epicycle is centered on the sun, although he did know that he had to keep the center of the Venusian epicycle aligned with the sun. In this sense, the two systems are not precisely equivalent; nonetheless, any motion in one of them can be replicated by the other.

Similarly for the outer planets. Let the configuration be:

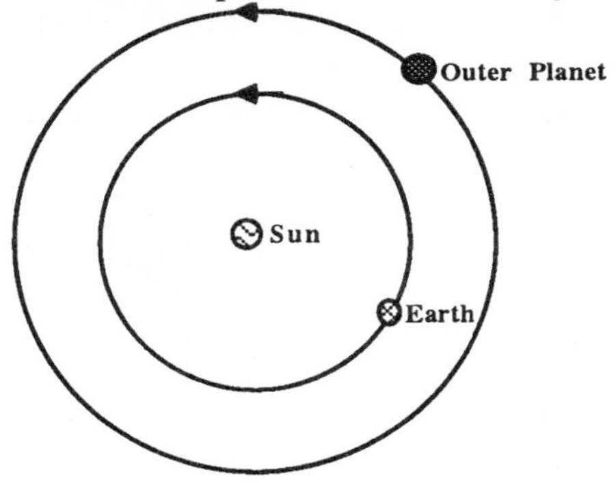


The motions of S and P can be represented in the following manner. First of all, S will move around E in a circular orbit of radius SE. To represent the motion of P, it is necessary to make its distance from E vary from SP – SE to SP + SE. This can be ensured by placing P on an epicycle of radius equal to SE, which epicycle moves on a deferent of radius equal to SP.

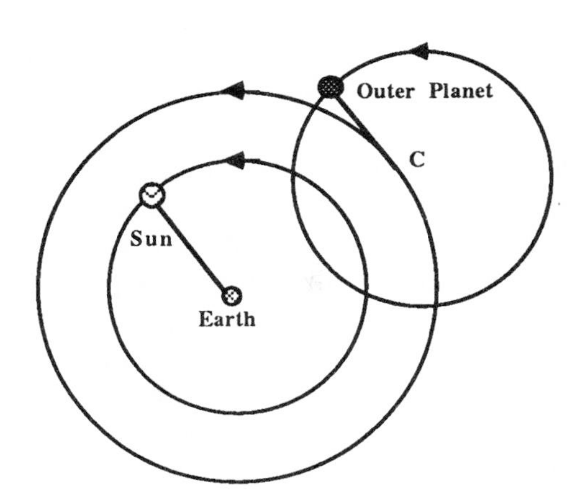

To ensure that retrograde motions of the outer planets will always occur when they are in opposition, it is necessary to stipulate that the line PC must remain parallel to SE. Ptolemy accomplished this by having the radius of the epicycles for each outer planet remain parallel to a line from the earth to the mean sun. Note that this also ensures that the distance SP will remain constant. Put differently, we can see from this diagram that the epicyclic motion of the outer planets is ultimately a projection of the earth's orbital motion. This also makes clear why Ptolemy had to give Mars a large epicycle. The reason is that compared to the radius of the Martian orbit, the radius of the earth's orbit is quite large. In fact, if the radius of the earth's orbit is 10, that of Mars is 15.2. We see that 10/15.2 is approximately equal to 39.5/60. That is, the ratio of the radius of the earth's orbit to that of the Martian orbit in the Copernican system nearly equals the ratio of the radius of Martian epicycle to the radius of its deferent circle in the Ptolemaic system. The same applies to the other outer planets as can be verified from data in the following table, which provides the modern values for the various parameters of the planets according to the Copernican theory.

| Planet | Orbital Radius | Sidereal Period | Synodic Period |
|---|---|---|---|
| Mercury | 3.87 | 87.97 days | 115.88 days |
| Venus | 7.23 | 224.70 days | 583.92 days |
| Earth | 10.00 | 1 yr. | |
| Mars | 15.23 | 1.881 yrs. | 779.94 days |
| Jupiter | 52.03 | 11.862 yrs. | 398.88 days |
| Saturn | 95.39 | 29.458 yrs. | 378.09 days |

In concluding this comparison of the Ptolemaic and Copernican orbital configurations, it is worth noting that the planets do not move precisely in circles, but rather in ellipses. Because, however, the eccentricities of the planetary ellipses are quite small, the orbits can to a good degree of accuracy be treated as if they were circles. Moreover, it would require only the employment of more elaborate, although essentially comparable, mathematics to show that any motion occurring in a heliocentric system can be accounted for within a geocentric system.

*Some Problems with the Copernican System*

Numerous problems beset the Copernican system. These can be classified under three headings.

(1) The Problem of Parallax and the Concomitant "Infinitization" of the Universe

If the earth moves around the sun, stars should appear to shift their positions. Such an effect is called **parallax**. Consider, for example, the case of stars located in the plane of the earth's orbit. This is represented in the next diagram.

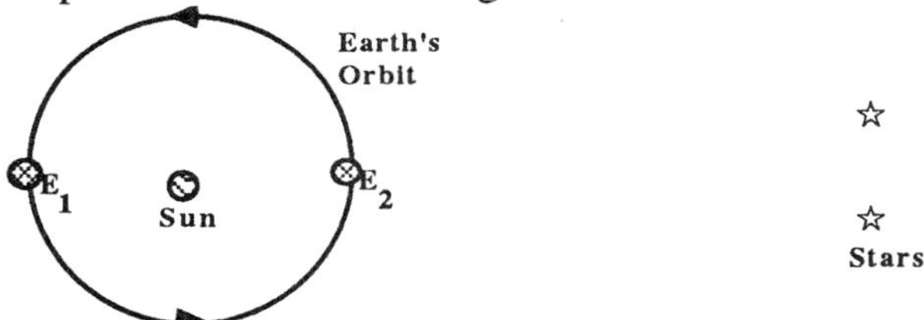


As the earth moves from $E_1$ to $E_2$, the two stars should appear to move apart, just as the headlights of a parked car separate as you approach them. The next diagram illustrates the situation for stars near the pole of the ecliptic.

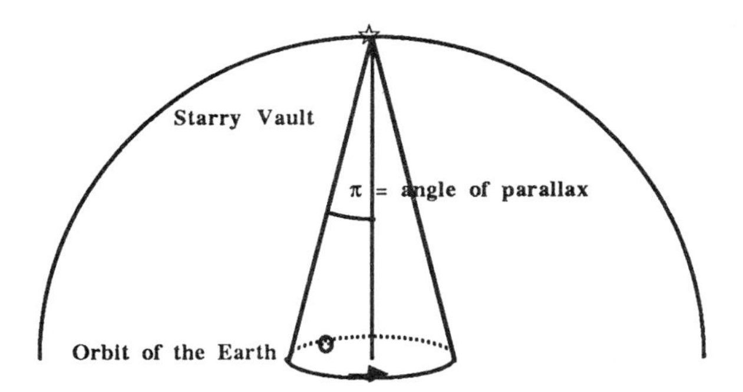


As the earth moves through its orbit, stars near the pole of the ecliptic appear to move in circles, whereas stars lower in the sky move in the form of an ellipse. The angle of parallax, $\pi$, is defined as the angle subtended by half of the semi-major axis of the earth's orbit. No such parallactic effects had ever been discovered. Copernicus tried to explain away this failure to find parallax by saying that the stars are so remote that the effect, although present, is beyond the limits of observation. In this sense, the Copernican system **"infinitized"** the universe or at least entailed an immense increase in its size. This in turn created problems as to what could be the function of the vast empty space between the solar system and the stars.

(2) Physical Problems

Copernicus also had to face severe physical problems because of the earth's rotation and revolution. Let us consider these quantitatively. The radius of the earth is about 4,000 miles as the Greeks knew; its circumference is about ($2 \times \pi \times 4{,}000$ miles =) 25,000 miles. Consequently, a person on the equator is whirled around at the daily rate of about (25,000 miles/24 hours =) 1000 m.p.h. due to the earth's rotation. Copernicus assigned the sun a distance of 1,142 earth radii from the earth, that is, approximately 4.6 million miles. The modern value for the earth-sun distance is far larger: 93 million miles. Using the Copernican value, we can calculate the average speed of the earth, assuming that it revolves around the sun. It is equal to $4.6 \times 10^6 \times 2\pi$ miles/($365 \times 24^h$) = about 3,200 m.p.h. The modern value is about 69,000 m.p.h. Such speeds, it would seem, would produce very obvious physical effects, such as clouds, birds, and projectiles being left behind as the earth moves. It was, then, a serious problem for the Copernicans to explain why such effects are not observed.

(3) Theological Problems

These were not only numerous, but also varied in nature. One problem was that the Copernican system seemed to contradict various scriptural passages. As an example, consider Joshua 10:13 in which God commanded the sun and moon to stand still. Persons pointed out that this passage would lose its meaning were the sun not in motion. Other scriptural passages occur in Ecclesiastes 1:4–5 and Psalm 92. Moreover, were the earth turned into just another planet and possibly the stars into suns with their own planetary systems, then the earth would seem to lose on the physical level the primacy that seems to be assigned to it by the Christian doctrine that the earth was the scene of Christ's incarnation and redemptive actions. One facet of this is that by the end of the sixteenth century, persons were actively speculating about extraterrestrial beings.

*Conclusion*

These Copernican materials along with the selection from his *De revolutionibus*, which follows, show some of the strengths and weaknesses of his system. It should be apparent that in a sense the choice between the two systems rested on aesthetic criteria. This correspondingly suggests why Copernicus urged that his system is more harmonious, more natural, and simpler than Ptolemy's. On the other hand, it also indicates why Galileo praised Copernicus for being able to commit rape on his senses.

## NICHOLAS COPERNICUS

# ON THE REVOLUTIONS[2]

NICHOLAS COPERNICUS OF TORUŃ
SIX BOOKS ON THE REVOLUTIONS OF THE HEAVENLY SPHERES

Diligent reader, in this work, which has just been created and published, you have the motions of the fixed stars and planets, as these motions have been reconstituted on the basis of ancient as well as recent observations, and have moreover been embellished by new and marvelous hypotheses. You also have most convenient tables, from which you will be able to compute those motions with the utmost ease for any time whatever. Therefore buy, read, and enjoy [this work].

*Let no one untrained in geometry enter here.*

NUREMBERG
JOHANNES PETREIUS
1543

[2]This reading consists of the introductory materials and Chapters 1–10 of Book One of *De revolutionibus orbium coelestium.* It is reprinted with the generous permission of the Polish Academy of Sciences from Nicholas Copernicus, *On the Revolutions,* edited by Jerzy Dobrzycki and translated with commentary by Edward Rosen (Baltimore: Johns Hopkins Univ. Press, 1978).

## To the Reader
## Concerning the Hypotheses of this Work[3]

There have already been widespread reports about the novel hypotheses of this work, which declares that the earth moves whereas the sun is at rest in the center of the universe. Hence certain scholars, I have no doubt, are deeply offended and believe that the liberal arts, which were established long ago on a sound basis, should not be thrown into confusion. But if these men are willing to examine the matter closely, they will find that the author of this work has done nothing blameworthy. For it is the duty of an astronomer to compose the history of the celestial motions through careful and expert study. Then he must conceive and devise the causes of these motions or hypotheses about them. Since he cannot in any way attain to the true causes, he will adopt whatever suppositions enable the motions to be computed correctly from the principles of geometry for the future as well as for the past. The present author has performed both of these duties excellently. For these hypotheses need not be true nor even probable. On the contrary, if they provide a calculus consistent with the observations, that alone is enough. Perhaps there is someone who is so ignorant of geometry and optics that he regards the epicycle of Venus as probable, or thinks that it is the reason why Venus sometimes precedes and sometimes follows the sun by forty degrees and even more. Is there anyone who is not aware that from this assumption it necessarily follows that the diameter of the planet at perigee should appear more than four times, and the body of the planet more than sixteen times, as great as at apogee? Yet this variation is refuted by the experience of every age. In this science there are some other no less important absurdities, which need not be set forth at the moment. For this art, it is quite clear, is completely and absolutely ignorant of the causes of the apparent nonuniform motions. And if any causes are

---

[3][It is important to be aware that this letter to the reader was written by Andreas Osiander (1498–1552), a Lutheran theologian, who had taken responsibility for seeing *De revolutionibus* through the press. It was not identified as by Osiander in the first edition of the book, nor, according to most scholars, would Copernicus have endorsed the claims made in it. MJC]

devised by the imagination, as indeed very many are, they are not put forward to convince anyone that they are true, but merely to provide a reliable basis for computation. However, since different hypotheses are sometimes offered for one and the same motion (for example, eccentricity and an epicycle for the sun's motion), the astronomer will take as his first choice that hypothesis which is the easiest to grasp. The philosopher will perhaps seek the semblance of the truth. But neither of them will understand or state anything certain, unless it has been divinely revealed to him.

Therefore alongside the ancient hypotheses, which are no more probable, let us permit these new hypotheses also to become known, especially since they are admirable as well as simple and bring with them a huge treasure of very skillful observations. So far as hypotheses are concerned, let no one expect anything certain from astronomy, which cannot furnish it, lest he accept as the truth ideas conceived for another purpose, and depart from this study a greater fool than when he entered it. Farewell.

## Letter of Nicholas Schönberg

*Nicholas Schönberg, Cardinal of Capua,*
*to Nicholas Copernicus, Greetings.*

Some years ago word reached me concerning your proficiency, of which everybody constantly spoke. At that time I began to have a very high regard for you, and also to congratulate our contemporaries among whom you enjoyed such great prestige. For I had learned that you had not merely mastered the discoveries of the ancient astronomers uncommonly well but had also formulated a new cosmology. In it you maintain that the earth moves; that the sun occupies the lowest, and thus the central, place in the universe; that the eighth heaven remains perpetually motionless and fixed; and that, together with the elements included in its sphere, the moon, situated between the heavens of Mars and Venus, revolves around the sun in the period of a year. I have also learned that you have written an exposition of this whole system of astronomy, and have computed the planetary motions and set them down in tables, to the greatest admiration of all.

Therefore with the utmost earnestness I entreat you, most learned sir, unless I inconvenience you, to communicate this discovery of yours to scholars, and at the earliest possible moment to send me your writings on the sphere of the universe together with the tables and whatever else you have that is relevant to this subject. Moreover, I have instructed Theodoric of Regen to have everything copied in your quarters at my expense and dispatched to me. If you gratify my desire in this manner, you will see that you are dealing with a man who is zealous for your reputation and eager to do justice to so fine a talent. Farewell.

Rome, 1 November 1536

## To His Holiness, Pope Paul III, Nicholas Copernicus' Preface to His Book On the Revolutions

I can readily imagine, Holy Father, that as soon as some people hear that in this volume, which I have written about the revolutions of the spheres of the universe, I ascribe certain motions to the terrestrial globe, they will shout that I must be immediately repudiated together with this belief. For I am not so enamored of my own opinions that I disregard what others may think of them. I am aware that a philosopher's ideas are not subject to the judgement of ordinary persons, because it is his endeavor to seek the truth in all things, to the extent permitted to human reason by God. Yet I hold that completely erroneous views should be shunned. Those who know that the consensus of many centuries has sanctioned the conception that the earth remains at rest in the middle of the heaven as its center would, I reflected, regard it as an insane pronouncement if I made the opposite assertion that the earth moves. Therefore I debated with myself for a long time whether to publish the volume which I wrote to prove the earth's motion or rather to follow the example of the Pythagoreans and certain others, who used to transmit philosophy's secrets only to kinsmen and friends, not in writing but by word of mouth, as is shown by Lysis' letter to Hipparchus. And they did so, it seems to me, not, as some suppose, because they were in some way jealous about their teachings, which would be spread around; on the very contrary, they wanted the very beautiful thoughts attained by great men of deep devotion not to be ridiculed by those who are reluctant to exert themselves vigorously in any literary pursuit unless it is lucrative; or if they are stimulated to the non-acquisitive study of philosophy by the exhortation and example of others, yet because of their dullness of mind they play the same part among philosophers as drones among bees. When I weighed these considerations, the scorn which I had reason to fear on account of the novelty and unconventionality of my opinion almost induced me to abandon completely the work which I had undertaken.

But while I hesitated for a long time and even resisted, my friends drew me back. Foremost among them was the cardinal of Capua, Nicholas Schönberg, renowned in every field of learning. Next to him was a man who loves me dearly, Tiedemann Giese, bishop of Chelmno, a close student of sacred letters as well as of all good literature. For he repeatedly encouraged me and, sometimes adding reproaches, urgently requested me to publish this volume and finally permit it to appear after being buried among my papers and lying concealed not merely until the ninth year but by now the fourth period of nine years. The same conduct was recommended to me by not a few other very eminent scholars. They exhorted me no longer to refuse, on account of the fear which I felt, to make my work available for the general use of students of astronomy. The crazier my doctrine of the earth's motion now appeared to most people the argument ran, so much the more admiration and thanks would it gain after they saw the publication of my writings dispel the fog of absurdity by most luminous proofs. Influenced therefore by these persuasive men and by this hope, in the end I allowed my friends to bring out an edition of the volume, as they had long besought me to do.

However, Your Holiness will perhaps not be greatly surprised that I have dared to publish my studies after devoting so much effort to working them out that I did not hesitate to put down my thoughts about the earth's motion in written form too. But you are rather waiting to hear from me how it occurred to me to venture to conceive any motion of the earth, against the traditional opinion of astronomers and almost against common sense. I have accordingly no desire to conceal from Your Holiness that I was impelled to consider a different system of deducing the motions of the universe's spheres for no other reason than the realization that astronomers do not agree among themselves in their investigations of this subject. For, in the first place, they are so uncertain about the motion of the sun and moon that they cannot establish and observe a constant length even for the tropical year. Secondly, in determining the motions not only of those bodies but also of the other five planets, they do not use the same principles, assumptions, and explanations of the apparent revolutions and motions. For while some employ only homocentrics, others utilize eccentrics and epicycles, and yet they do not quite reach their goal. For although those who put their faith in homocentrics showed that some nonuniform motions could be compounded in this way, nevertheless by this means they were unable to obtain any incontrovertible results in absolute agreement with the phenomena. On the other hand, those who devised the eccentrics seem thereby in large measure to have solved the problem of the apparent motions with appropriate calculations. But meanwhile they introduced a good many ideas which apparently contradict the first principles of uniform motion. Nor could they elicit or deduce from the eccentrics the principal consideration, that is, the structure of the universe and the true symmetry of its parts. On the contrary, their experience was just like some one taking from various places hands, feet, a head, and other pieces, very well depicted, it may be, but not for the representation of a single person; since these fragments would not belong to one another at all, a monster rather than man would be put together from them. Hence in the process of demonstration or "method", as it is called, those who employed eccentrics are found either to have omitted something essential or to have admitted something extraneous and wholly irrelevant. This would not have happened to them, had they followed sound principles. For if the hypotheses assumed by them were not false, everything which follows from their hypotheses would be confirmed beyond doubt. Even though what I am now saying may be obscure, it will nevertheless become clearer in the proper place.

For a long time, then, I reflected on this confusion in the astronomical traditions concerning the derivation of the motions of the universe's spheres. I began to be annoyed that the movements of the world machine, created for our sake by the best and most systematic Artisan of all, were not understood with greater certainty by the philosophers, who otherwise examined so precisely the most insignificant trifles of this world. For this reason I undertook the task of rereading the works of all the philosophers which I could obtain to learn whether anyone had ever proposed other motions of the universe's spheres than those expounded by the teachers of astronomy in the schools. And in fact first I found in Cicero that Hicetas supposed the earth to move. Later I also discovered in Plutarch that certain others were of this opinion. I have decided to set his words down here, so that they may be available to everybody:

> Some think that the earth remains at rest. But Philolaus the Pythagorean believes that, like the sun and moon, it revolves around the fire in an oblique circle. Heraclides of Pontus and Ecphantus the Pythagorean make the earth move, not in a progressive motion, but like a wheel in a rotation from west to east about its own center.

Therefore, having obtained the opportunity from these sources, I too began to consider the mobility of the earth. And even though the idea seemed absurd, nevertheless I knew that others before me had been granted the freedom to imagine any circles whatever for the purpose of explaining the heavenly phenomena. Hence I thought that I too would be readily permitted to ascertain whether explanations sounder than those of my predecessors could be found for the revolution of the celestial spheres on the assumption of some motion of the earth.

Having thus assumed the motions which I ascribe to the earth later on in the volume, by long and intense study I finally found that if the motions of the other planets are correlated with the orbiting of the earth, and are computed for the revolution of each planet, not only do their phenomena follow therefrom but also the order and size of all the planets and spheres, and heaven itself is so linked together that in no portion of it can anything be shifted without disrupting the remaining parts and the universe as a whole. Accordingly in the arrangement of the volume too I have adopted the following order. In the first book I set forth the entire distribution of the spheres together with the motion which I attribute to the earth, so that this book contains, as it were, the general structure of the universe. Then in the remaining books I correlate the motions of the other planets and of all the spheres with the movement of the earth so that it may thereby determine to what extent the motions and appearances of the other planets and spheres can be saved if they are correlated with the earth's motions. I have no doubt that acute and learned astronomers will agree with me if, as this discipline especially requires, they are willing to examine and consider, not superficially but thoroughly, what I adduce in this volume in the proof of these matters. However, in order that the educated and uneducated alike may see that I do not run away from the judgement of anybody at all, I have preferred dedicating my studies to Your Holiness rather than to anyone else. For even in this very remote corner of the earth where I live you are considered the highest authority by virtue of the loftiness of your office and your love for all literature and astronomy too. Hence by your prestige and judgement you can easily suppress calumnious attacks although, as the proverb has it, there is no remedy for a backbite.

Perhaps there will be blabbers who claim to be judges of astronomy although completely ignorant of the subject and, badly distorting some passage of Scripture to their purpose, will dare to find fault with my undertaking and censure it. I disregard them even to the extent of despising their criticism as unfounded. For it is not unknown that Lactantius, otherwise an illustrious writer but hardly an astronomer, speaks quite childishly about the earth's shape, when he mocks those who declare that the earth has the form of a globe. Hence scholars need not be surprised if any such persons will likewise ridicule me. Astronomy is written for astronomers. To them my work too will seem, unless I am mistaken, to make some contribution also to the Church, at the head of which Your Holiness now stands. For not so long ago under Leo X the Lateran Council considered the problem of reforming the ecclesiastical calendar. The issue remained undecided then only because the length of the year and month and the motions of the sun and moon were regarded as not yet adequately measured. From that time on, at the suggestion of that most distinguished man, Paul, bishop of Fossombrone, who was then in charge of this matter, I have directed my attention to a more precise study of the topics. But what I have accomplished in this regard, I leave to the judgement of Your Holiness in particular and of all other learned astronomers. And lest I appear to Your Holiness to promise more about the usefulness of this volume than I can fulfill, I now turn to the work itself.

## BOOK ONE

### INTRODUCTION

Among the many various literary and artistic pursuits which invigorate men's minds, the strongest affection and utmost zeal should, I think, promote the studies concerned with the most beautiful objects, most deserving to be known. This is the nature of the discipline which deals with the universe's divine revolutions, the asters' motions, sizes, distances, risings and settings, as well as the causes of the other phenomena in the sky, and which, in short, explains its whole appearance. What indeed is more beautiful than heaven, which of course contains all things of beauty? This is proclaimed by its very names [in Latin] *caelum* and *mundus*, the latter denoting purity and ornament, the former a carving. On account of heaven's transcendent perfection most philosophers have called it a visible god. If then the value of the arts is judged by the subject matter which they treat, that art will be by far the foremost which is labeled astronomy by some, astrology by others, but by many of the ancients, the consummation of mathematics. Unquestionably the summit of the liberal arts and most worthy of a free man, it is supported by almost all the branches of mathematics. Arithmetic, geometry, optics, surveying, mechanics and whatever others there are all contribute to it.

Although all the good arts serve to draw man's mind away from vices and lead it toward better things, this function can be more fully performed by this art, which also provides extraordinary intellectual pleasure. For when a man is occupied with things which he sees established in the finest order and directed by divine management, will not the unremitting contemplation of them and a certain familiarity with them stimulate him to the best and to admiration for the Maker of everything, in whom are all happiness and every good? For would not the godly Psalmist [92:4] in vain declare that he was made glad through the work of the Lord and rejoiced in the works of His hands, were not we drawn to the contemplation of the highest good by this means, as though by a chariot?

The great benefit and adornment which this art confers on the commonwealth (not to mention the countless advantages to individuals) are most excellently observed by Plato. In the *Laws*, Book VII, he thinks that it should be cultivated chiefly because by dividing time into groups of days as months and years, it would keep the state alert and attentive to the festival and sacrifices. Whoever denies its necessity for the teacher of any branch of higher learning is thinking foolishly, according to Plato. In his opinion it is highly unlikely that anyone lacking the requisite knowledge of the sun, moon, and other heavenly bodies can become and be called godlike.

However, this divine rather than human science, which investigates the loftiest subjects, is not free from perplexities. The main reason is that its principles and assumptions, called "hypotheses" by the Greeks, have been a source of disagreement, as we see, among most of those who undertook to deal with this subject, and so they did not rely on the same ideas. An additional reason is that the motion of the planets and the revolution of the stars could not be measured with numerical precision and completely understood except with the passage of time and aid of many earlier observations, through which this knowledge was transmitted to posterity from hand to hand, so to say. To be sure, Claudius Ptolemy of Alexandria, who far excels the rest by his wonderful skill and industry, brought this entire art almost to perfection with the help of observations extending over a period of more than four hundred years, so that there no longer seemed to be any gap which he had not closed. Nevertheless very many things, as we perceive, do not agree with the conclusion which ought to follow from his system, and besides certain other motions have been discovered which were not yet known to him. Hence Plutarch too, in discussing the sun's tropical year, says that so far the motion of the heavenly bodies has eluded the skill of the astronomers. For, to use the year itself as an example, it is well known, I think, how different the opinions concerning it have always been, so that many have abandoned all hope that an exact determination of it could be found. The situation is the same with regard to other heavenly bodies.

Nevertheless, to avoid giving the impression that this difficulty is an excuse for indolence, by a grace of God, without whom we can accomplish nothing, I shall attempt a broader inquiry into these matters. For, the number of aids we have to assist our enterprise grows with the interval of time extending from the originators of this art to us. Their discoveries may be compared with what I have newly found. I acknowledge, moreover, that I shall treat many topics differently from my predecessors, and yet I shall do so thanks to them, for it was they who first opened the road to the investigation of these very questions.

## Chapter 1. THE UNIVERSE IS SPHERICAL

First of all, we must note that the universe is spherical. The reason is either that, of all forms, the sphere is the most perfect, needing no joint and being a complete whole, which can be neither increased nor diminished; or that it is the most capacious of figures, best suited to enclose and retain all things; or even that all the separate parts of the universe, I mean the sun, moon, planets and stars, are seen to be of this shape; or that wholes strive to be circumscribed by this boundary, as is apparent in drops of water and other fluid bodies when they seek to be self-contained. Hence no one will question the attribution of this form to the divine bodies.

## Chapter 2. THE EARTH TOO IS SPHERICAL

The earth also is spherical, since it presses upon its center from every direction. Yet it is not immediately recognized as a perfect sphere on account of the great height of the mountains and depth of the valleys. They scarcely alter the general sphericity of the earth, however, as is clear from the following considerations. For a traveler going from any place toward the north, that pole of the daily rotation gradually climbs higher, while the opposite pole drops down an equal amount. More stars in the north are seen not to set, while in the south certain stars are no longer seen to rise. Thus Italy does not see Canopus, which is visible in Egypt; and Italy does see the River's last star, which is unfamiliar to our area in the colder region. Such stars, conversely, move higher in the heavens for a traveller heading southward, while those which are high in our sky sink down. Meanwhile, moreover, the elevations of the poles have the same ratio everywhere to the portions of the earth that have been traversed. This happens on no other figure than the sphere. Hence the earth too is evidently enclosed between poles and is therefore spherical. Furthermore, evening eclipses of the sun and moon are not seen by easterners, nor morning eclipses by westerners, while those occurring in between are seen later by easterners but earlier by westerners.

The waters press down into the same figure also, as sailors are aware, since land which is not seen from a ship is visible from the top of its mast. On the other hand, if a light is attached to the top of the mast, as the ship draws away from the land, those who remain ashore see the light drop down gradually until it finally disappears, as though setting. Water, furthermore, being fluid by nature, manifestly always seeks the same lower levels as earth and pushes up from the shore no higher than its rise permits. Hence whatever land emerges out of the ocean is admittedly that much higher.

## Chapter 3. HOW EARTH FORMS A SINGLE SPHERE WITH WATER

Pouring forth its seas everywhere, then, the ocean envelops the earth and fills its deeper chasms. Both tend toward the same center because of their heaviness. Accordingly there had to be less water than land, to avoid having the water engulf the entire earth and to have the water recede from some portions of the land and from the many islands lying here and there, for the preservation of living creatures. For what are the inhabited countries and the mainland itself but an island larger than the others?

We should not heed certain peripatetics who declared that the entire body of water is ten times greater than all the land. For, according to the conjecture which they accepted, in the transmutation of the elements as one unit of earth dissolves, it becomes ten units of water. They also assert that the earth bulges out to some extent as it does because it is not of equal weight everywhere on account of its cavities, its center of gravity being different from its center of magnitude. But they err through ignorance of the art of geometry. For they do not realize that the water cannot be even seven times greater and still leave any part of the land dry, unless earth as a whole vacated the center of gravity and yielded that position to water, as if the latter were heavier than itself. For spheres are to each other as the cubes of their diameters. Therefore, if earth were the eighth part of seven parts of water, earth's diameter could not be greater than the distance from [their joint] center to the circumference of the waters. So far are they from being as much as ten times greater [than the land].

Moreover, there is no difference between the earth's centers of gravity and magnitude. This can be established by the fact that from the ocean inward the curvature of the land does not mount steadily in a continuous rise. If it did, it would keep the sea water out completely and in no way permit the inland seas and such vast gulfs to intrude. Furthermore, the depth of the abyss would never stop increasing from the shore of the ocean outward, so that no island or reef or any form of land would be encountered by sailors on the longer voyages. But it is well known that almost in the middle of the inhabited lands barely fifteen furlongs remain between the eastern Mediterranean and the Red Sea. On the other hand, in his *Geography* Ptolemy extended the habitable area halfway around the world. Beyond that meridian, where he left unknown land, the moderns have added Cathay and territory as vast as sixty degrees of longitude, so that now the earth is inhabited over a greater stretch of longitude than is left for the ocean. To these regions, moreover, should be added the islands discovered in our time under the rulers of Spain and Portugal, and especially America, named after the ship's captain who found it. On account of its still undisclosed size it is thought to be a second group of inhabited countries. There are also many other islands, heretofore unknown. So little reason have we to marvel at the existence of antipodes or antichthones. Indeed, geometrical reasoning about the location of America compels us to believe that it is diametrically opposite the Ganges district of India.

From all these facts, finally, I think it is clear that land and water together press upon a single center of gravity; that the earth has no other center of magnitude; that, since earth is heavier, its gaps are filled with water; and that consequently there is little water in comparison with land, even though more water perhaps appears on the surface.

The earth together with its surrounding waters must in fact

have such a shape as its shadow reveals, for it eclipses the moon with the arc of a perfect circle. Therefore the earth is not flat, as Empedocles and Anaximenes thought; nor drum-shaped, as Leucippus; nor bowl-shaped, as Heraclitus; nor hollow in another way, as Democritus; nor again cylindrical, as Anaximander; nor does its lower side extend infinitely downward, the thickness diminishing toward the bottom, as Xenophanes taught; but it is perfectly round, as the philosophers hold.

## Chapter 4. THE MOTION OF THE HEAVENLY BODIES IS UNIFORM, ETERNAL, AND CIRCULAR OR COMPOUNDED OF CIRCULAR MOTIONS

I shall now recall to mind that the motion of the heavenly bodies is circular, since the motion appropriate to a sphere is rotation in a circle. By this very act the sphere expresses its form as the simplest body, wherein neither beginning nor end can be found, nor can the one be distinguished from the other, while the sphere itself traverses the same points to return upon itself.

In connection with the numerous [celestial] spheres, however, there are many motions. The most conspicuous of all is the daily rotation, which the Greeks call *nuchthemeron,* that is, the interval of a day and a night. The entire universe, with the exception of the earth, is conceived as whirling from east to west in this rotation. It is recognized as the common measure of all motions, since we even compute time itself chiefly by the number of days.

Secondly, we see other revolutions as advancing in the opposite direction, that is, from west to east; I refer to those of the sun, moon, and five planets. The sun thus regulates the year for us, and the moon the month, which are also very familiar periods of time. In like manner each of the other five planets completes its own orbit.

Yet [these motions] differ in many ways [from the daily rotation or first motion]. In the first place, they do not swing around the same poles as the first motion, but run obliquely through the zodiac. Secondly, these bodies are not seen moving uniformly in their orbits, since the sun and moon are observed to be sometimes slow, at other times faster in their course. Moreover, we see the other five planets also retrograde at times, and stationary at either end [of the regression]. And whereas the sun always advances along its own direct path, they wander in various ways, straying sometimes to the south and sometimes to the north; that is why they are called "planets" [wanderers]. Furthermore, they are at times nearer to the earth, when they are said to be in perigee; at other times they are farther away, when they are said to be in apogee.

We must acknowledge, nevertheless, that their motions are circular or compounded of several circles, because these nonuniformities recur regularly according to a constant law. This could not happen unless the motions were circular, since only the circle can bring back the past. Thus, for example, by a composite motion of circles the sun restores to us the inequality of days and nights as well as the four seasons of the year. Several motions are discerned herein, because a simple heavenly body cannot be moved by a single sphere nonuniformly. For this nonuniformity would have to be caused either by an inconstancy, whether imposed from without or generated from within, in the moving force or by an alteration in the revolving body. From either alternative, however, the intellect shrinks. It is improper to conceive any such defect in objects constituted in the best order.

It stands to reason, therefore, that their uniform motions appear nonuniform to us. The cause may be either that their circles have poles different [from the earth's] or that the earth is not at the center of the circles on which they revolve. To us who watch the course of these planets from the earth, it happens that our eye does not keep the same distance from every part of the orbits, but on account of their varying distances these bodies seem larger when nearer than when farther away (as has been proved in optics). Likewise, in equal arcs of their orbits their motions will appear unequal in equal times on account of the observer's varying distance. Hence I deem it above all necessary that we should carefully scrutinize the relation of the earth to the heavens lest, in our desire to examine the loftiest objects, we remain ignorant of things nearest to us, and by the same error attribute to the celestial bodies what belongs to the earth.

## Chapter 5. DOES CIRCULAR MOTION SUIT THE EARTH? WHAT IS ITS POSITION?

Now that the earth too has been shown to have the form of a sphere, we must in my opinion see whether also in this case the form entails the motion, and what place in the universe is occupied by the earth. Without the answers to these questions it is impossible to find the correct explanation of what is seen in the heavens. To be sure, there is general agreement among the authorities that the earth is at rest in the middle of the universe. They hold the contrary view to be inconceivable or downright silly. Nevertheless, if we examine the matter more carefully, we shall see that this problem has not yet been solved, and is therefore by no means to be disregarded.

Every observed change of place is caused by a motion of either the observed object or the observer or, of course, by an unequal displacement of each. For when things move with equal speed in the same direction, the motion is not perceived, as between the observed object and the observer, I mean. It is the earth, however, from which the celestial ballet is beheld in its repeated performances before our eyes. Therefore, if any motion is ascribed to the earth, in all things outside it the same motion will appear, but in the opposite direction, as though they were moving past it. Such in particular is the daily rotation, since it seems to involve the entire universe except the earth and what is around it. However, if you grant that the heavens have no part in this motion but that the earth rotates from west to east, upon earnest consideration you will find that this is the actual situation concerning the apparent rising and setting of the sun, moon, stars and planets. Moreover since the heavens, which enclose and provide the setting for everything, constitute the space common to all things, it is not at first blush clear why motion should not be attributed rather to the enclosed than to the enclosing, to the thing located in space rather than to the framework of space. This opinion was indeed maintained by Heraclides and Ecphantus, the Pythagoreans, and by Hicetas of Syracuse, according to Cicero. They rotated the earth in the middle of the universe, for they ascribed the setting of the stars to the earth's interposition, and their rising to its withdrawal.

If we assume its daily rotation, another and no less important question follows concerning the earth's position. To be sure, heretofore there has been virtually unanimous acceptance of the belief that the middle of the universe is the earth. Anyone who denies that the earth occupies the middle or center of the universe may nevertheless assert that its distance [therefrom] is insignificant in comparison with [the distance of] the sphere of the fixed stars, but perceptible and noteworthy in relation to the spheres of the sun and the other planets. He may deem this to be the reason why their motions appear nonuniform, as conforming to a center other than the center of the earth. Perhaps he can [thereby] produce a not inept explanation of the apparent nonuniform motion. For the fact that the same planets are observed nearer to the earth and farther away necessarily proves that the center of the earth is not the center of their circles. It is less clear whether the approach and withdrawal are executed by the earth or the planets.

It will occasion no surprise if, in addition to the daily rotation, some other motion is assigned to the earth. That the earth rotates, that it also travels with several motions, and that it is one of the heavenly bodies are said to have been the opinions of Philolaus the Pythagorean. He was no ordinary astronomer, inasmuch as Plato did not delay going to Italy for the sake of visiting him, as Plato's biographers report.

But many have thought it possible to prove by geometrical reasoning that the earth is in the middle of the universe; that being like a point in relation to the immense heavens, it serves as their center; and that it is motionless because, when the universe moves, the center remains unmoved, and the things nearest to the center are carried most slowly.

## Chapter 6. THE IMMENSITY OF THE HEAVENS COMPARED TO THE SIZE OF THE EARTH

The massive bulk of the earth does indeed shrink to insignificance in comparison with the size of the heavens. This can be ascertained from the fact that the boundary circles (for that is the translation of the Greek term *horizons*) bisect the entire sphere of the heavens. This could not happen if the earth's size or distance from the universe's center were noteworthy in comparison with the heavens. For, a circle that bisects a sphere passes through its center, and is the greatest circle that can be described on it.

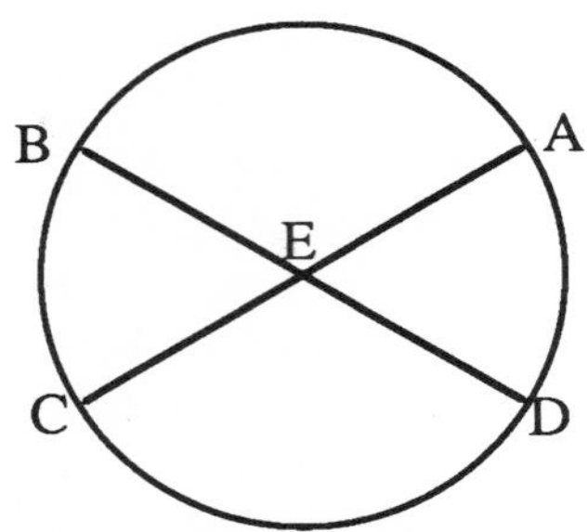


Thus let circle ABCD be a horizon, and let the earth, from which we do our observing be E, the center of the horizon, which separates what is seen from what is not seen. Now, through a dioptra or horoscopic instrument or water level placed at E, let the first point of the Crab be sighted rising at point C, and at that instant the first point of the Goat is perceived to be setting at A. Then A, E, and C are on a straight line through the dioptra. This line is evidently a diameter of the ecliptic, since six visible signs form a semicircle, and E, the [line's] center, is identical with the horizon's center. Again, let the signs shift their position until the first point of the Goat rises at B. At that time the Crab will also be observed setting at D. BED will be a straight line and a diameter of the ecliptic. But, as we have already seen, AEC also is a diameter of the same circle. Its center, obviously, is the intersection [of the diameters]. A horizon, then, in this way always bisects the ecliptic, which is a great circle of the sphere. But on a sphere, if a circle bisects any great circle, the bisection circle is itself a great circle. Consequently a horizon is one of the great circles, and its center is clearly identical with the center of the ecliptic.

Yet a line drawn from the earth's surface [to a point in the firmament] must be distinct from the line drawn from the earth's center [to the same point]. Nevertheless, because these lines are immense in relation to the earth, they become like parallel lines [III, 15]. Because their terminus is enormously remote they appear to be a single line. For in comparison with their length the space enclosed by them becomes imperceptible, as is demonstrated in optics. This reasoning certainly makes it quite clear that the heavens are immense by comparison with the earth and present the aspect of an infinite magnitude, while on the testimony of the senses the earth is related to the heavens as a point to a body, and a finite to an infinite magnitude.

But no other conclusion seems to have been established. For it does not follow that the earth must be at rest in the middle of the universe. Indeed, a rotation in twenty-four hours of the enormously vast universe should astonish us even more than a rotation of its least part, which is the earth. For, the argument that the center is motionless, and what is nearest the center moves the least, does not prove that the earth is at rest in the middle of the universe.

To take a similar case, suppose you say that the heavens rotate but the poles are stationary, and what is closest to the poles moves the least. The Little Bear, for example, being very close to the pole, is observed to move much more slowly than the Eagle or the Little Dog because it describes a smaller circle. Yet all these constellations belong to a single sphere. A sphere's movement, vanishing at its axis, does not permit an equal motion of all its parts. Nevertheless these are brought round in equal times, though not over equal spaces, by the rotation of the whole sphere. The upshot of the argument, then, is the claim that the earth as a part of the celestial sphere shares in the same nature and movement so that, being close to the center, it has a slight motion. Therefore, being a body and not the center, it too will describe arcs like those of a celestial circle, though smaller, in the same time. The falsity of this contention is clearer than daylight. For it would always have to be noon in one place, and always midnight in another, so that the daily risings and settings could not take place, since the motion of the whole and the part would be one and inseparable.

But things separated by the diversity of their situations are subject to a very different relation: those enclosed in a smaller orbit revolve faster than those traversing a bigger circle. Thus Saturn, the highest of the planets, revolves in thirty years; the moon, undoubtedly the nearest to the earth, completes its course in a month; and to close the series, it will be thought, the earth rotates in the period of a day and a night. Accordingly the same question about the daily rotation emerges again. On the other hand, likewise still undetermined is the earth's position, which has been made even less certain by what was said above. For that proof establishes no conclusion other than the heavens' unlimited size in relation to the earth. Yet how far this immensity extends is not at all clear. At the opposite extreme are the very tiny indivisible bodies called "atoms". Being imperceptible, they do not immediately constitute a visible body when they are taken two or a few at a time. But they can be multiplied to such an extent that in the end there are enough of them to combine in a perceptible magnitude. The same may be said also about the position of the earth. Although it is not in the center of the universe, nevertheless its distance therefrom is still insignificant, especially in relation to the sphere of the fixed stars.

## Chapter 7. WHY THE ANCIENTS THOUGHT THAT THE EARTH REMAINED AT REST IN THE MIDDLE OF THE UNIVERSE AS ITS CENTER

Accordingly, the ancient philosophers sought to establish that the earth remains at rest in the middle of the universe by certain other arguments. As their main reason, however, they adduce heaviness and lightness. Earth is in fact the heaviest element, and everything that has weight is borne toward it in an effort to reach its inmost center. The earth being spherical, by their own nature heavy objects are carried to it from all directions at right angles to its surface. Hence, if they were not checked at its surface, they would collide at its center, since a straight line perpendicular to a horizontal plane at its point of tangency with a sphere leads to the [sphere's] center. But things brought to the middle, it seems to follow, come to rest at the middle. All the more, then, will the entire earth be at rest in the middle, and as the recipient of every falling body it will remain motionless thanks to its weight.

In like manner, the ancient philosophers analyze motion and its nature in a further attempt to confirm their conclusion. Thus, according to Aristotle, the motion of a single simple body is simple; of the simple motions, one is straight and the other is circular; of the straight motions, one is upward and the other is downward. Hence every simple motion is either toward the middle, that is, downward; or away from the middle, that is, upward; or around the middle, that is, circular. To be carried downward, that is, to seek the middle, is a property only of earth and water, which are considered heavy; on the other hand, air and fire, which are endowed with lightness, move upward and away from the middle. To these four elements it seems reasonable to assign rectilinear motion, but to the heavenly bodies, circular motion around the middle. This is what Aristotle says [*Heavens*, I, 2; II, 14].

Therefore, remarks Ptolemy of Alexandria [*Syntaxis*, I, 7], if the earth were to move, merely in a daily rotation, the opposite of what was said above would have to occur, since a motion would have to be exceedingly violent and its speed unsurpassable to carry the entire circumference of the earth around in twenty-four hours. But things which undergo an abrupt rotation seem utterly unsuited to gather [bodies to themselves], and seem more likely, if they have been produced by combination, to fly apart unless they are held together by some bond. The earth would long ago have burst asunder, he says, and dropped out of the skies (a quite preposterous notion); and, what is more, living creatures and any other loose weights would by no means remain unshaken. Nor would objects falling in a straight line descend perpendicularly to their appointed place, which would meantime have been withdrawn by so rapid a movement. Moreover, clouds and anything else floating in the air would be seen drifting always westward.

## Chapter 8. THE INADEQUACY OF THE PREVIOUS ARGUMENTS AND A REFUTATION OF THEM

For these and similar reasons forsooth the ancients insist that the earth remains at rest in the middle of the universe, and that this is its status beyond any doubt. Yet if anyone believes that the earth rotates, surely he will hold that its motion is natural, not violent. But what is in accordance with nature produces effects contrary to those resulting from violence, since things to which force or violence is applied must disintegrate and cannot long endure. On the other hand, that which is brought into existence by nature is well-ordered and preserved in its best state. Ptolemy has no cause, then, to fear that the earth and everything earthly will be disrupted by a rotation created through nature's handiwork, which is quite different from what art or human intelligence can accomplish.

But why does he not feel this apprehension even more for the universe, whose motion must be the swifter, the bigger the heavens are than the earth? Or have the heavens become immense because the indescribable violence of their motion drives them away from the center? Would they also fall apart if they came to a halt? Were this reasoning sound, surely the size of the heavens would likewise grow to infinity. For the higher they are driven by the power of their motion, the faster that motion will be, since the circumference of which it must make the circuit in the period of twenty-four hours is constantly expanding; and, in turn, as the velocity of the motion mounts, the vastness of the heavens is enlarged. In this way the speed will increase the size, and the size the speed, to infinity. Yet according to the familiar axiom of physics that the infinite cannot be traversed or moved in any way, the heavens will therefore necessarily remain stationary.

But beyond the heavens there is said to be no body, no space, no void, absolutely nothing, so that there is nowhere the heavens can go. In that case it is really astonishing if something can be held in check by nothing. If the heavens are infinite, however, and finite at their inner concavity only, there will perhaps be more reason to believe that beyond the heavens there is nothing. For, every single thing, no matter what size it attains, will be inside them, but the heavens will abide motionless. For, the chief contention by which it is sought to prove that the universe is finite is its motion. Let us therefore leave the question whether the universe is finite or infinite to be discussed by the natural philosophers.

We regard it as a certainty that the earth, enclosed between poles, is bounded by a spherical surface. Why then do we still hesitate to grant it the motion appropriate by nature to its form rather than attribute a movement to the entire universe, whose limit is unknown and unknowable? Why should we not admit, with regard to the daily rotation, that the appearance is in the heavens and the reality in the earth? This situation closely resembles what Vergil's Aeneas says: "Forth from the harbor we sail, and the land and the cities slip backward [*Aeneid*, III, 72]." For when a ship is floating calmly along, the sailors see its motion mirrored in everything outside, while on the other hand they suppose that they are stationary, together with everything on board. In the same way, the motion of the earth can unquestionably produce the impression that the entire universe is rotating.

Then what about the clouds and the other things that hang in the air in any manner whatsoever, or the bodies that fall down, and conversely those that rise aloft? We would only say that not merely the earth and the watery element joined with it have this motion, but also no small part of the air and whatever is linked in the same way to the earth. The reason may be either that the nearby air, mingling with earthy or watery matter, conforms to the same nature as the earth, or that the air's motion, acquired from the earth by proximity, shares without resistance in its unceasing rotation. No less astonishingly, on the other hand, is the celestial movement declared to be accompanied by the uppermost belt of air. This is indicated by those bodies that appear suddenly, I mean, those that the Greeks called "comets" and "bearded stars". Like the other heavenly bodies, they rise and set. They are thought to be generated in that region. That part of the air, we can maintain, is unaffected by the earth's motion on account of its great distance from the earth. The air closest to the earth will accordingly seem to be still. And so will the things suspended in it, unless they are tossed to and fro, as indeed they are, by the wind or some other disturbance. For what else is the wind in the air but the wave in the sea?

We must in fact avow that the motion of falling and rising bodies in the framework of the universe is twofold, being in every case a compound of straight and circular. For, things that sink of their own weight, being predominantly earthy, undoubtedly retain the same nature as the whole of which they are parts. Nor is the explanation different in the case of those things, which, being fiery, are driven forcibly upward. For also fire here on the earth feeds mainly on earthy matter, and flame is defined as nothing but blazing smoke. Now it is a property of fire to expand what it enters. It does this with such great force that it cannot be prevented in any way by any device from bursting through restraints and completing its work. But the motion of expansion is directed from the center to the circumference. Therefore, if any part of the earth is set afire, it is carried from the middle upwards. Hence the statement that the motion of a simple body is simple holds true in particular for circular motion, as long as the simple body abides in its natural place and with its whole. For when it is in place, it has none but circular motion, which remains wholly within itself like a body at rest. Rectilinear motion, however, affects things which leave their natural place or are thrust out of it or quit it in any manner whatsoever. Yet nothing is so incompatible with the orderly arrangement of the universe and the design of the totality as something out of place. Therefore rectilinear motion occurs only to things that are not in proper condition and are not in complete accord with their nature, when they are separated from their whole and forsake its unity.

Furthermore, bodies that are carried upward and downward, even when deprived of circular motion, do not execute a simple, constant, and uniform motion. For they cannot be governed by their lightness or by the impetus of their weight. Whatever falls moves slowly at first, but increases its speed as it drops. On the other hand, we see this earthly fire (for we behold no other), after it has been lifted up high, slacken all at once, thereby revealing the reason to be the violence applied to the earthy matter. Circular motion, however, always rolls along uniformly, since it has an unfailing cause. But rectilinear motion has a cause that quickly stops functioning. For when rectilinear motion brings bodies to their own place, they cease to be heavy or light, and their motion ends. Hence, since circular motion belongs to wholes, but parts have rectilinear motion in addition, we can say that "circular" subsists with "rectilinear" as "being alive" with "being sick". Surely Aristotle's division of simple motion into three types, away from the middle, toward the middle, and around the middle, will be construed merely as a logical exercise. In like manner we distinguish line, point, and surface, even though one cannot exist without another, and none of them without body.

As a quality, moreover, immobility is deemed nobler and more divine than change and instability, which are therefore better suited to the earth than to the universe. Besides, it would seem quite absurd to attribute motion to the framework of space or that which encloses the whole of space, and not, more appropriately, to that which is enclosed and occupies some space, namely, the earth.

Last of all, the planets obviously approach closer to the earth and recede farther from it. Then the motion of a single body around the middle, which is thought to be the center of the earth, will be both away from the middle and also toward it. Motion around the middle, consequently, must be interpreted in a more general way, the sufficient condition being that each such motion encircle its own center. You see, then, that all these arguments make it more likely that the earth moves than that it is at rest. This is especially true of the daily rotation, as particularly appropriate to the earth. This is enough, in my opinion, about the first part of the question.

## Chapter 9. CAN SEVERAL MOTIONS BE ATTRIBUTED TO THE EARTH? THE CENTER OF THE UNIVERSE

Accordingly, since nothing prevents the earth from moving, I suggest that we should now consider also whether several motions suit it, so that it can be regarded as one of the planets. For, it is not the center of all the revolutions. This is indicated by the planets' apparent nonuniform motion and their varying distances from the earth. These phenomena cannot be explained by circles concentric with the earth. Therefore, since there are many centers, it will not be by accident that the further question arises whether the center of the universe is identical with the center of terrestrial gravity or with some other point. For my part I believe that gravity is nothing but a certain natural desire, which the divine providence of the Creator of all things has implanted in parts, to gather as a unity and a whole by combining in the form of a globe. This impulse is present, we may suppose, also in the sun, the moon, and the other brilliant planets, so that through its operation they remain in that spherical shape which they display. Nevertheless, they swing round their circuits in divers ways. If, then, the earth too moves in other ways, for example, about a center, its additional motions must likewise be reflected in many bodies outside it. Among these motions we find the yearly revolution. For if this is transformed from a solar to a terrestrial movement, with the sun acknowledged to be at rest, the risings and settings which bring the zodiacal signs and fixed stars into view morning and evening will appear in the same way. The stations of the planets, moreover, as well as their retrogradations and [resumption of] forward motion will be recognized as being, not movements of the planets, but a motion of the earth, which the planets borrow for their own appearances. Lastly, it will be realized that the sun occupies the middle of the universe. All these facts are disclosed to us by the principle governing the order in which the planets follow one another, and by the harmony of the entire universe, if only we look at the matter, as the saying goes, with both eyes.

## Chapter 10. THE ORDER OF THE HEAVENLY SPHERES

Of all things visible, the highest is the heaven of the fixed stars. This, I see, is doubted by nobody. But the ancient philosophers wanted to arrange the planets in accordance with the duration of the revolutions. Their principle assumes that of objects moving equally fast, those farther away seem to travel more slowly, as is proved in Euclid's *Optics*. The moon revolves in the shortest period of time because, in their opinion, it runs on the smallest circle as the nearest to the earth. The highest planet, on the other hand, is Saturn, which completes the biggest circuit in the longest time. Below it is Jupiter, followed by Mars.

With regard to Venus and Mercury, however, differences of opinion are found. For, these planets do not pass through every elongation from the sun, as the other planets do. Hence Venus and Mercury are located above the sun by some authorities, like Plato's *Timaeus* [38D], but below the sun by others, like Ptolemy [*Syntaxis*, IX, 1] and many of the moderns. Al-Bitruji places Venus above the sun, and Mercury below it.

According to Plato's followers, all the planets, being dark bodies otherwise, shine because they receive sunlight. If they were below the sun, therefore, they would undergo no great elongation from it, and hence they would be seen halved or at any rate less than fully round. For, the light which they receive would be reflected mostly upward, that is, toward the sun, as we see in the new or dying moon. In addition, they argue, the sun must sometimes be eclipsed by the interposition of these planets, and its light cut off in proportion to their size. Since this is never observed, these planets do not pass beneath the sun at all, according to those who follow Plato.

On the other hand, those who locate Venus and Mercury below the sun base their reasoning on the wide space which they notice between the sun and the moon. For the moon's greatest distance from the earth is 64 1/6 earth-radii. This is contained, according to them, about 18 times in the sun's least distance from the earth, which is 1160 earth-radii. Therefore between the sun and the moon there are 1096 earth-radii [≈1160 – 64 1/6]. Consequently, to avoid having so vast a space remain empty, they announce that the same numbers almost exactly fill up the apsidal distances, by which they compute the thickness of those spheres. Thus the moon's apogee is followed by Mercury's perigee. Mercury's apogee is succeeded by the perigee of Venus, whose apogee, finally, almost reaches the sun's perigee. For between the apsides of Mercury they calculate about 177 1/2 earth-radii. Then the remaining space is very nearly filled by Venus' interval of 910 earth-radii.

Therefore they do not admit that these heavenly bodies have any opacity like the moon's. On the contrary, these shine either with their own light or with the sunlight absorbed throughout their bodies. Moreover, they do not eclipse the sun, because it rarely happens that they interfere with our view of the sun, since they generally deviate in latitude. Besides, they are tiny bodies in comparison with the sun. Venus, although bigger than Mercury, can occult barely a hundredth of the sun. So says Al-Battani of Raqqa, who thinks that the sun's diameter is ten times larger [than Venus'], and therefore so minute a speck is not easily descried in the most brilliant light. Yet in his *Paraphrase* of Ptolemy, Ibn Rushd reports having seen something blackish when he found a conjunction of the sun and Mercury indicated in the tables. And thus these two planets are judged to be moving below the sun's sphere.

But this reasoning also is weak and unreliable. This is obvious from the fact that there are 38 earth-radii to the moon's perigee, according to Ptolemy [*Syntaxis*, V, 13], but more than 49 according to a more accurate determination, as will be made clear below. Yet so great a space contains, as we know, nothing but air and, if you please, also what is called "the element of fire". Moreover, the diameter of Venus' epicycle which carries it $45^{\circ}$ more or less to either side of the sun, must be six times longer than the line drawn from the earth's center to Venus' perigee, as will be demonstrated in the proper place [V, 21]. In this entire space which would be taken up by that huge epicycle of Venus and which, moreover, is so much bigger than what would accommodate the earth, air, aether, moon, and Mercury, what will they say is contained if Venus revolved around a motionless earth?

Ptolemy [*Syntaxis*, IX, 1] argues also that the sun must move in the middle between the planets which show every elongation from it and those which do not. This argument carries no conviction because its error is revealed by the fact that the moon too shows every elongation from the sun.

Now there are those who locate Venus and then Mercury below the sun, or separate these planets [from the sun] in some other sequence. What reason will they adduce to explain why Venus and Mercury do not likewise traverse separate orbits divergent from the sun, like the other planets, without violating the arrangement [of the planets] in accordance with their [relative] swiftness and slowness? Then one of two alternatives will have to be true. Either the earth is not the center to which the order of the planets and spheres is referred, or there really is no principle of arrangement nor any apparent reason why the highest place belongs to Saturn rather than to Jupiter or any other planet.

In my judgement, therefore, we should not in the least disregard what was familiar to Martianus Capella, the author of an encyclopedia, and to certain other Latin writers. For according to them, Venus and Mercury revolve around the sun as their center. This is the reason, in their opinion, why these planets diverge no farther from the sun than is permitted by the curvature of their revolutions. For they do not encircle the earth, like the other planets, but "have opposite circles". Then what else do these authors mean but that the center of their spheres is near the sun? Thus Mercury's sphere will surely be enclosed within Venus', which by common consent is more than twice as big, and inside that wide region it will occupy a space adequate for itself. If anyone seizes this opportunity to link Saturn, Jupiter, and Mars also to that center, provided he understands their spheres to be so large that together with Venus and Mercury the earth too is enclosed inside and encircled, he will not be mistaken, as is shown by the regular pattern of their motions.

For [these outer planets] are always closest to the earth, as is well known, about the time of their evening rising, that is, when they are in opposition to the sun, with the earth between them and the sun. On the other hand, they are at their farthest from the earth at the time of their evening setting, when they become invisible in the vicinity of the sun, namely, when we have the sun between them and the earth. These facts are enough to show that their center belongs more to the sun, and is identical with the center around which Venus and Mercury likewise execute their revolutions.

But since all these planets are related to a single center, the space remaining between Venus' convex sphere and Mars' concave sphere must be set apart as also a sphere or spherical shell, both of whose surfaces are concentric with those spheres. This [intercalated sphere] receives the earth together with its attendant, the moon, and whatever is contained within the moon's sphere. Mainly for the reason that in this space we find quite an appropriate and adequate place for the moon, we can by no means detach it from the earth, since it is incontrovertibly nearest to the earth.

Hence I feel no shame in asserting that this whole region engirdled by the moon, and the center of the earth, traverse this grand circle amid the rest of the planets in an annual revolution around the sun. Near the sun is the center of the universe. Moreover, since the sun remains stationary, whatever appears as a motion of the sun is really due rather to the motion of the earth. In comparison with any other spheres of the planets, the distance from the earth to the sun has a magnitude which is quite appreciable in proportion to those dimensions. But the size of the universe is so great that the distance earth-sun is imperceptible in relation to the sphere of the fixed stars. This should be admitted, I believe, in preference to perplexing the mind with an almost infinite multitude of spheres, as must be done by those who kept the earth in the middle of the universe. On the contrary, we should rather heed the wisdom of nature. Just as it especially avoids producing anything superfluous or useless, so it frequently prefers to endow a single thing with many effects.

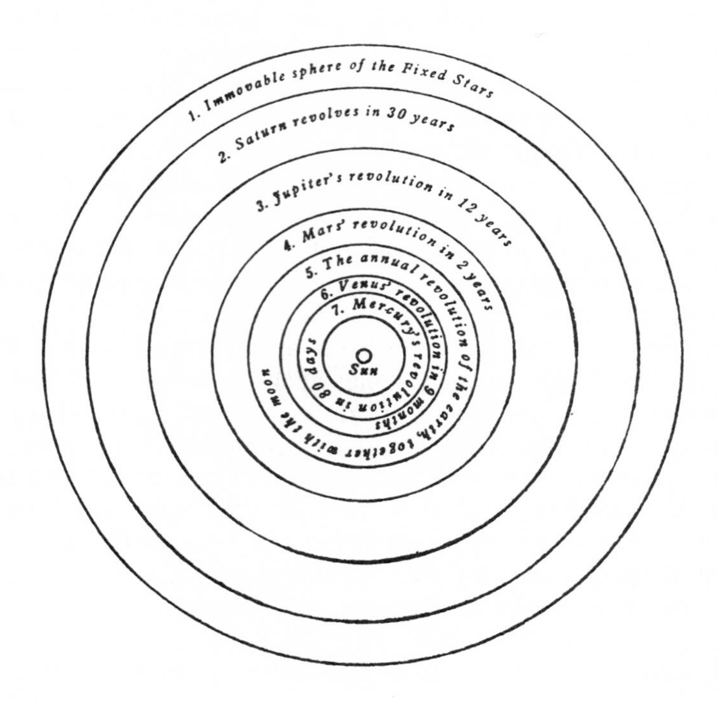


All these statements are difficult and almost inconceivable, being of course opposed to the beliefs of many people. Yet, as we proceed, with God's help I shall make them clearer than sunlight, at any rate to those who are not unacquainted with the science of astronomy. Consequently, with the first principle remaining intact, for nobody will propound a more suitable principle than that the size of the spheres is measured by the length of the time, the order of the spheres is the following, beginning with the highest.

The first and the highest of all is the sphere of the fixed stars, which contains itself and everything, and is therefore immovable. It is unquestionably the place of the universe, to which the motion and position of all the other heavenly bodies are compared. Some people think that it also shifts in some way. A different explanation of why this appears to be so will be adduced in my discussion of the earth's motion [I, 11].

[The sphere of the fixed stars] is followed by the first of the planets, Saturn, which completes its circuit in 30 years. After

Saturn, Jupiter accomplishes its revolution in 12 years. Then Mars revolves in 2 years. The annual revolution takes the series' fourth place, which contains the earth, as I said [earlier in I, 10], together with the lunar sphere as an epicycle. In the fifth place Venus returns in 9 months. Lastly, the sixth place is held by Mercury, which revolves in a period of 80 days.

At rest, however, in the middle of everything is the sun. For in this most beautiful temple, who would place this lamp in another or better position than that from which it can light up the whole thing at the same time? For, the sun is not inappropriately called by some people the lantern of the universe, its mind by others, and its ruler by still others. [Hermes] the Thrice Greatest labels it a visible god, and Sophocles' Electra, the all-seeing. Thus indeed, as though seated on a royal throne, the sun governs the family of planets revolving around it. Moreover, the earth is not deprived of the moon's attendance. On the contrary, as Aristotle says in a work on animals, the moon has the closest kinship with the earth. Meanwhile the earth has intercourse with the sun, and is impregnated for its yearly parturition.

In this arrangement, therefore, we discover a marvelous symmetry of the universe, and an established harmonious linkage between the motion of the spheres and their size, such as can be found in no other way. For this permits a not inattentive student to perceive why the forward and backward arcs appear greater in Jupiter than in Saturn and smaller than in Mars, and on the other hand greater in Venus than in Mercury. This reversal in direction appears more frequently in Saturn than in Jupiter, and also more rarely in Mars and Venus than in Mercury. Moreover, when Saturn, Jupiter, and Mars rise at sunset, they are nearer to the earth than when they set in the evening or appear at a later hour. But Mars in particular, when it shines all night, seems to equal Jupiter in size, being distinguished only by its reddish color. Yet in the other configurations it is found barely among the stars of the second magnitude, being recognized by those who track it with assiduous observations. All these phenomena proceed from the same cause, which is in the earth's motion.

Yet none of these phenomena appears in the fixed stars. This proves their immense height, which makes even the sphere of the annual motion, or its reflection, vanish from before our eyes.

For, every visible object has some measure of distance beyond which it is no longer seen, as is demonstrated in optics. From Saturn, the highest of the planets, to the sphere of the fixed stars there is an additional gap of the largest size. This is shown by the twinkling lights of the stars. By this token in particular they are distinguished from the planets, for there had to be a very great difference between what moves and what does not move. So vast, without any question, is the divine handiwork of the most excellent Almighty.

## *Some Self-Testing Questions Concerning Copernicus and His System*

1. Did Copernicus discover his new system as a result of new observations?

2. Did Copernicus use new and improved mathematical techniques?

3. Did Copernicus have access to and/or make use of instruments superior to those available to Ptolemy?

4. Is the Copernican system in principle capable of greater accuracy than the Ptolemaic system?

5. Was the Copernican system in fact more accurate than the Ptolemaic system?

6. Are the arguments of Copernicus more scientific and more free of philosophical encumbrances and presuppositions than those of Ptolemy?

7. Was Copernicus in opposition to the Church?

8. Is Copernicus's system simpler than Ptolemy's?

9. Is Copernicus's system more harmonious than Ptolemy's?

10. Was the Copernican system contradicted by any observations?

11. Which system, the Copernican or Ptolemaic, presented the larger number of problems for traditional physical and theological thought?

*Problems*

1. According to the Copernican system, if the radius of the orbit of the earth is 10, the radius of the orbit of Venus is 7.23. Use this information to estimate the radius of the Venusian epicycle in the Ptolemaic system, taking its deferent radius as 60. Check your answer by comparing it to the figure assigned to the Venusian epicycle in Chapter Four.

2. In the Copernican system, the radius of the orbit of Jupiter is 52.03 for the radius of the earth's orbit being set at 10. Use this data to calculate the radius of Jupiter's epicycle in the Ptolemaic system. Check your result by comparing it with the figure for Jupiter's epicycle in the table of epicycle and deferent radii in the chapter on the Ptolemaic system.

3. Given that the synodic period of Jupiter is 398.88 days, determine the average length of time between its retrogradations and also calculate Jupiter's sidereal period. Compare your result to the figure for the sidereal period of Jupiter.

4. Copernicans frequently stressed that their system made it possible to dispense with the large Ptolemaic epicycles for Mars, Jupiter, and Saturn. Explain why this was a feature of the Copernican system.

5. Copernicus specified that the average distance between the earth and sun is 1,142 earth radii, which figure is quite close to the value Ptolemy favored. The earth's radius is about 4,000 miles. At the time of Copernicus, some astronomers thought they could measure angles as small as 1 minute of arc. Given these figures and the fact that no one was able to measure a stellar parallax, i.e., a shift in the position of stars due to the earth's revolution around the sun, determine the least distance at which the stars can be positioned from the earth. Express the answer in terms of earth radii so as to provide a clear comparison with the Ptolemaic system. Hint: The value of sin 1' is .000291.

6. It is helpful in getting an idea of the relative size of the Copernican planetary orbits to plot the orbits on a map of the continental United States. Placing the sun at Chicago and letting the distance between the sun and Mercury be 100 miles, determine where on this scale the other planets should be placed.

# Chapter Seven

## *The Tychonic System*

**Tycho Brahe**

Who is the greatest astronomer alive today? Had that question been asked in 1600, the response would almost certainly have been **Tycho Brahe** (1546–1601). Given this fact, it is interesting to ask whether Brahe favored the Ptolemaic or the Copernican system. The answer is neither, but rather an ingenious system of his own devising. This will be discussed shortly, but first let us survey his life to see how he attained such a position of eminence.

*Life of Brahe*

Tycho Brahe was born in Denmark in 1546 to a family of minor nobility. In 1559 (aged 13), he entered the University of Copenhagen where his talents soon became evident. His interest in astronomy is usually dated from 1560 when he witnessed a solar eclipse. Ths eclipse, and especially the successful prediction of its occurrence by astronomers, deeply impressed him. In 1562, he proceeded to the University of Leipzig, the plan being for him to study law. His increasing fascination with astronomy, however, led him to read all the books on this topic that he could secure. He was troubled at this time by learning that the thirteenth-century Alphonsine tables of planetary motions were off by as much as a month and that the Prutenic tables of 1551 were off by a few days.

During the 1560s, Brahe traveled in various cities of Europe, in one of which (Rostock) he lost his nose in a duel, in another of which (Augsburg) he acquired a nineteen-foot quadrant for astronomical observations. In 1571, because of his father's death, he returned to Denmark to take up residence there. In 1572, he noticed a new star, a **nova** as such objects are called, in the constellation Cassiopeia. Attaining a brightness rivaling Venus, this star attracted widespread attention as did a short treatise on it that Brahe published in 1573. This nova raised problems for the Aristotelian cosmology according to which the heavens are unchanging. In 1574, he began to give lectures on astronomy at the University of Copenhagen and also to seek financial support for the founding of an observatory. Funding came from Frederick II, the Danish king, who gave him a small island, Hven, off the coast of Copenhagen. The cornerstone of his observatory and house, Uraniborg (see next picture), was laid in 1576 and developed into a sort of celestial palace with such elaborate instrumentation that Brahe estimated that all of this cost the king over a ton of gold. A second, smaller observatory, Stjerneborg, was also erected on the island, making simultaneous, independent observation possible.

**Uraniborg**

Another feature of Uraniborg was Brahe's giant **mural quadrant** (see next picture), used to determine the height above the horizon of celestial objects and the time at which they crossed the meridian. From such observations, he could determine their exact positions.

With his splendid (pre-telescopic) instruments, Brahe was able to attain an **unprecedented accuracy** of within one or two minutes of arc. Late in 1577, he observed a **comet**, his observations permitting him to prove that it was at least six times farther from the earth than the moon. This raised problems for the Aristotelian system, which portrayed comets as sublunar. His method in making this determination was to seek a parallax for the comet. He also, throughout his life, attempted to find a stellar parallax, but he did not succeed in this endeavor. In 1588, Brahe published a book on the comet of 1577, his *De mundi aetherei recentioribus phaenomenis*, in which he not only discussed that comet, but also set out a new system of the celestial motions, which has come to be known as the Tychonic system. It provided a third theory of the planetary system, to rival the Ptolemaic and Copernican systems.

**Brahe's Giant Mural Quadrant**
(Brahe is pictured in the background)

*The Tychonic System*

In setting the stage for his system, Brahe criticized both the Ptolemaic and the Copernican systems. He faulted the Ptolemaic system for its use of the equant as well as for its lack of elegance in accounting for the stationary points and retrogressions of the planets. He also protested against what he described as "that newly introduced innovation of the great Copernicus," who, Brahe admitted, had introduced devices "by which he very elegantly obviates those things which occur superfluously and incongruously in the Ptolemaic system, and does not at all offend against mathematical principles." In opposition to Copernicus, Brahe stated:

> . . . the body of the Earth, large, sluggish and inapt for motion is not to be disturbed by movement . . . any more than the Aetherial Lights [stars] are to be shifted, so that such ideas are opposed to physical principles and also the authority of Holy Writ. . . . Consequently I shall not speak now of the vast space between the orb of Saturn and the Eighth Sphere [the starry vault] left utterly empty of stars by this reasoning. . . .

This led him to the "opinion, beyond all possible doubt, that the Earth . . . occupies the center of the universe. . . ."

Brahe's criticisms of the Copernican system can be classified under three headings.

(1) It violates physical principles.
(2) It necessitates a stellar parallax.
(3) It contradicts Holy Writ.

The implications that Brahe drew from the second point are particularly noteworthy. The failure to find stellar parallax, he urged, would entail, if the Copernican theory were true, that the stars must be located extremely far beyond Saturn. This in turn would entail the absurd notion of a vast and useless vacant space between Saturn's orbit and the starry vault. Although Brahe's three objections were all serious, it is noteworthy that he attributed a greater elegance to the Copernican system and admitted that it "does not at all offend against mathematical principles."

Brahe's system is represented in the next diagram, which is based on one of his own representations of his system. In it, Saturn, Jupiter, and Mars revolve counterclockwise in large circular orbits centered on the sun. Venus and Mercury also orbit the sun, moving counterclockwise on smaller circles centered on the sun rather than on the earth. The sun revolves counterclockwise on a circle centered on the earth, carrying the orbits of Saturn, Jupiter, Mars, Venus, and Mercury with it. Because the earth is envisioned as being motionless (no revolution or rotation) in the center of the system, the entire arrangement, including the starry vault, must rotate once each day so as to account for the daily motion of the celestial bodies. Brahe believed his system to be free of the difficulties that he had detected in the Ptolemaic and in the Copernican system.

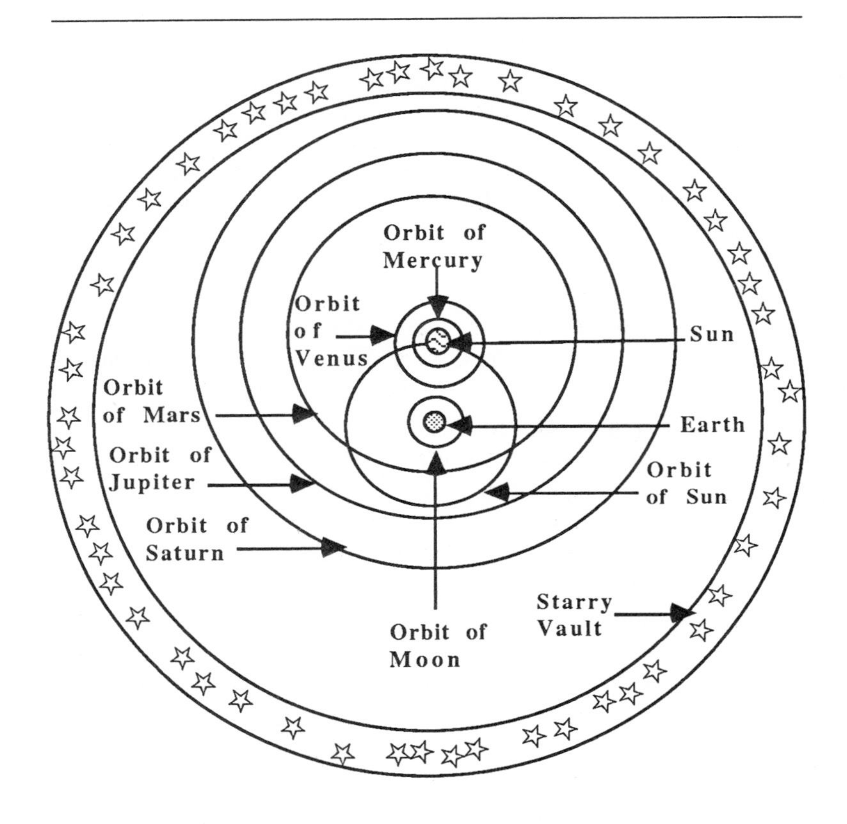


**Diagram of the Tychonic System**

A good test of whether one understands the Tychonic system consists in attempting to answer the following questions.

1. How are the bounded elongations of Venus and Mercury explained in it?
2. An inspection of his diagram shows that the solar and Martian orbits cross. Does this mean that those two objects can collide?
3. How could Brahe account for the retrogressions of the planets? In particular, would he do this in the same way as Ptolemy or as Copernicus?
4. Could Brahe, using this system, represent the celestial motions more accurately than Ptolemy or Copernicus had?
5. Are there conceivable astronomical observations that would refute this system?

6. How does the Tychonic system compare with the Ptolemaic and Copernican systems in explanatory power?

Brahe promised in his 1588 book that in a subsequent work he would present his system in full detail, with the appropriate epicycles, etc., and their parameters provided. He died without accomplishing this, his efforts having gone largely to making ever more precise observations. Shortly before he died, he hired the young **Johannes Kepler** to work with him, hoping that the more mathematically astute Kepler would work out the details of the Tychonic system for him.

Because of failing support from the Danish king, especially after the death of Frederick II in 1588, Brahe left Denmark in 1597, becoming in 1599 Imperial Mathematician to the court of Emperor Rudolph II in Prague. It was here that Kepler came to work for Brahe, succeeding him as Imperial Mathematician after Brahe's death in 1601.

### *Conclusion*

Tycho Brahe aspired to be remembered above all as a great cosmologist. In fact, his fame was largely due to his having been the finest pre-telescopic astronomical observer of all time. One evidence of the justice of such a description of him is that in plotting the positions of 21 key reference stars, he attained an accuracy of $\pm$40" of arc. His determinations of the planetary positions were nearly as accurate and provided Kepler with invaluable data for his later researches, especially because Brahe had observed the planets throughout their entire orbits rather than only at select positions as had been the previous practice. Although the Tychonic system is now regarded as little more than an interesting historical curiosity, considerable interest in it arose in astronomical circles during the last decade of the sixteenth century and early decades of the seventeenth century. This was because it represented in a number of ways an attractive alternative, which retained fundamental features of the Ptolemaic system while incorporating some aspects of the Copernican system.

*Appendix*
*Demonstration That the Ptolemaic and Tychonic Systems Generate the Same Motions for an Outer Planet*

**Method:** We shall start by taking a specific configuration of the sun, moon, and outer planet. From this we can get (part one) values for the relative distances involved in the two systems. We shall then (part two) set the objects in motion and show that after an arbitrary time t, the outer planet will be found in the same position in both systems.

**Part One:** Let us take the case where the outer planet is in conjunction with the sun. For this configuration, the planet will appear in the two systems as represented in the next diagram. We set $R_1$, the radius of the sun's orbit in the Tychonic system, as equal to the radius $R_1$ of the epicycle in the Ptolemaic system. We set the distance of the planet from the sun in the Tychonic system as equal to $R_2$ and set the radius of the Ptolemaic deferent also as $R_2$. Clearly for this situation, the systems are identical.

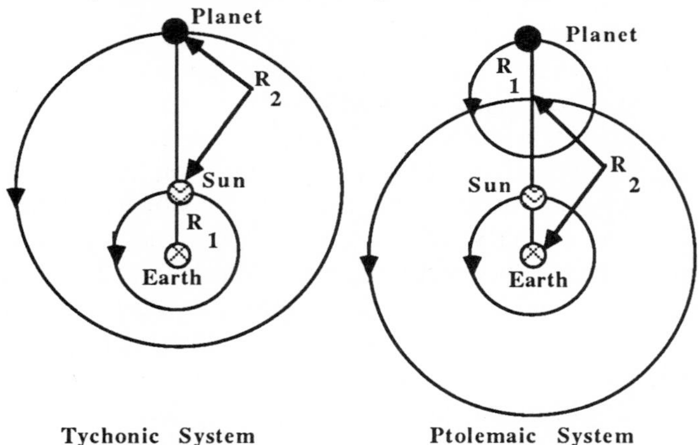


**Part Two:** Let us now determine what the configuration of the earth, sun, and planet will be after some arbitrary time t in each system. In other words, we shall let an identical period of time t elapse in each system and determine the relative positions of earth, sun, and planet. In time t, the planet will have moved in each system to a new location, $P_1$, and the sun will have moved to a new position, $S_1$. See the next diagram. We must prove that relative locations of $P_1$, $S_1$, and E will be the same in the two systems. We shall now calculate where $P_1$ will be seen for each system. For the Tychonic system, we stipulate that the planet moves around the sun at the rate of 360°/T where T is the sidereal period of the planet. Consequently, it will move in time t through $\angle ß$ where ß equals 360° x t/ T and where ß is measured from the apogee (top point) of the sun's orbit, which is itself moving. For the Ptolemaic system, we specify that the center of the epicycle D moves at the rate 360°/T with relation not to the sun but to the earth. Consequently, at the end of time t, D will also have moved through an angle ß = 360° x t/ T. And the planet will move on the epicycle in such a manner that the radius of the epicycle remains parallel to the line from the earth to the sun.

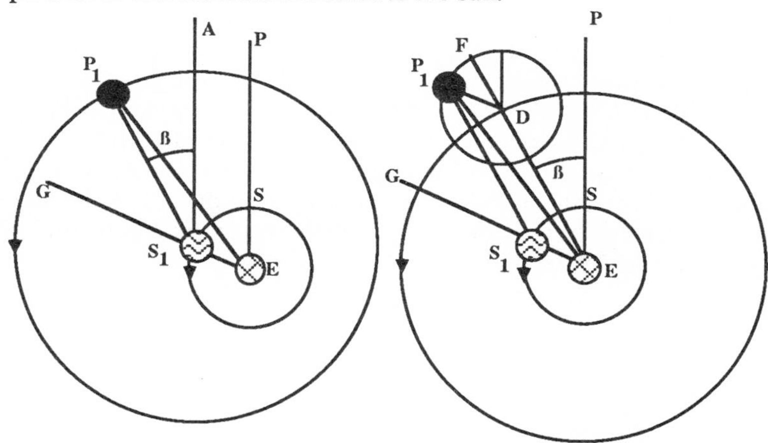


Tychonic System

Ptolemaic System

Let us now compare the positions of the earth, sun, and planet in each system. It is clear that the positions of $S_1$ and E relative to each other will be identical in the two systems and that, consequently, $\angle SES_1$ will be the same for each system. It is also clear that the length $S_1E$ will be the same in each of the two systems. We begin by determining the size of $\angle P_1S_1E$ for each system. For both systems, we extend a line from E through $S_1$ to some point G.

For the Tychonic system, it is evident that $\angle P_1S_1E = 180° - \angle P_1S_1G$. Letting A be the moving apogee of the planet's orbit, we have that because $AS_1$ is parallel to PE, it must be the case that $\angle AS_1G = \angle SES_1$. Consequently, $\angle P_1S_1E = 180° - \angle P_1S_1G = 180° - (\angle SES_1 - \angle ß)$.

For the Ptolemaic system, we also have that $\angle P_1S_1E = 180° - \angle P_1S_1G$. The conditions set down previously for the Ptolemaic system entail that D, the center of the epicycle, moves through $\angle\beta$ in time t. Because $P_1D = S_1E$ (as stated previously) and because these lines are also parallel (recall the Ptolemaic requirement that the radius of an outer planet's epicycle must remain parallel to a line from the earth to the sun), $S_1EDP_1$ is a parallelogram. Consequently, $\angle GS_1P_1 = \angle GED = \angle FDP_1$ where F is the apogee of the epicycle. But $\angle FDP_1 = \angle SES_1 - \angle\beta$. Consequently, again $\angle P_1S_1E = 180° - (\angle SES_1 - \angle\beta)$. Hence, $\angle P_1S_1E$ will be identical in the two systems.

All that is needed to complete the proof is to show that the Tychonic and Ptolemaic triangles $P_1S_1E$ are identical. Because in the Ptolemaic diagram $S_1EDP_1$ is a parallelogram, $P_1S_1 = DE$. But we know that DE is equal to $R_2$, as specified earlier. Because $P_1S_1$ (Tychonic) also equals $R_2$, we have $P_1S_1$(Ptolemaic) = $P_1S_1$ (Tychonic). It has already been shown that $S_1E$ (Tychonic) = $S_1E$ (Ptolemaic). Consequently, triangles $P_1S_1E$ are identical. Thus for any value of t, the two systems will position the earth, sun, and planet in identical locations. To put the point differently, any motion that can be accounted for by one of the systems can also be accounted for by the other.

# Chapter Eight

## *Johannes Kepler*

**Johannes Kepler**

### *Life of Kepler*

In his *History of Astronomy from Thales to Kepler*, J. L. E. Dreyer prefaced his discussion of Kepler by stating:

> In January, 1599, Mästlin, having heard from his former pupil, Johann Kepler, of the difficulties which Tycho Brahe had encountered in determining the excentricities of the

> planets, wrote in reply that Tycho had hardly left a shadow of what had hitherto been taken for astronomical science, and that only one thing was certain, which was that mankind knew nothing of astronomical matters.

With this thought in mind, let us turn to Johannes Kepler (1571–1630), who was born on December 27, 1571, to a ne'er-do-well father who served as a military mercenary for some years before abandoning his family in 1588, and to a mother who around 1620 was tried for being a witch. The boy was precocious above all in illness, being beset by smallpox, headaches, boils, rashes, worms, piles, the mange, and worst of all for an aspiring astronomer, defective eyesight. His visual problems included double vision in one eye and myopia in both eyes. Nonetheless, before his death in 1630, he had attained a substantial measure of fame. The following two descriptions of him, both from his own hand, vividly set out the polarities between which his self-image fluctuated and raise the question: which is nearer the truth? The first description is from a document composed around 1596, just as his first book was being published. In it, he described himself as having

> . . . in every respect a canine nature. He has the appearance of a delicate domestic puppy. 1. His body is agile, thin, well proportioned. Similarly in regard to nourishment; he enjoys gnawing bones, hard crusts of bread, is voracious without limit, and grabs whatever the eye observes. He drinks little. Is content with the very least. 2. Similarly for his habits. He perpetually (like a dog) insinuates himself with his superiors, depends on others for everything, ministers to them, never becomes angry with them when reproached, and in every way is eager to win back their favor. He probes everything in all the disciplines, in politics, in domestic matters, even the worst type. He is constantly in motion, following anyone in anything, following someone, imitating him in action and thought.
>
> He is impatient with conversations and often greets persons coming into a room, not unlike a dog. When the least thing is snatched away from him, he growls and gets aroused like a dog. He is a tenacious pursuer of the wicked, namely, he barks at them. And he bites with sarcastic remarks. To

> many people he is exceedingly hostile and they shun him, but his masters favor him, not unlike a good family dog. Like a dog, he has a horror of baths, tinctures, and lotions. His rashness is without restraint, as is surely due to Mars being in quadrature with Mercury and in trine with the moon. . . .
>
> He was accordingly praised by his teachers in his youth for his good nature, even though he was morally the worst among his peers. . . . When as a boy of ten, he was first able to read scripture . . . he was saddened that because of the impurity of his life, the honor of being a prophet was denied him. . . .[1]

The above passage, which reads almost as if it were from Dostoyevsky's fictional *Notes from the Underground*, contrasts sharply with Kepler's description of himself in the opening paragraph of his *Harmonices mundi* (1619) as having written a book "to be read either now or by posterity, it matters not. It can wait a century for a reader, as God himself has waited six thousand years for a witness." As we now encounter Kepler, it is useful to ask: which of these characterizations better captures the character and accomplishments of Kepler?

In 1589, Kepler entered the Lutheran university in Tübingen, where he learned astronomy from **Michael Maestlin,** who was sufficiently interested in the Copernican system that he introduced the young and gifted Kepler to its complexities. Kepler did well at Tübingen but left there in 1594 (with plans to return to take a position teaching mathematics at the Protestant seminary in Graz. While at Graz, he began to ponder various questions.

(1) Why are there six planets?
(2) Why are their orbits positioned as they are?
(3) Why do planets farther from the sun move more slowly?

On July 9, 1595, while teaching a class, an answer to the first two questions flashed across his mind. Referring to this discovery, Kepler attributed it to "Divine Providence," adding: "I believe this all the more because I have constantly prayed to God that I might succeed if what Copernicus had said was true." What happened was that Kepler was explaining to his class a pattern in the

[1]Johannes Kepler, *Gesammelte Werke,* vol. 19 (Munich, 1975), p. 336.

conjunctions of Jupiter and Saturn. Each such conjunction occurs eight zodiacal signs distant from the previous one and at a temporal spacing of about twenty years. This may be seen in the next diagram, which is from Kepler's writings.

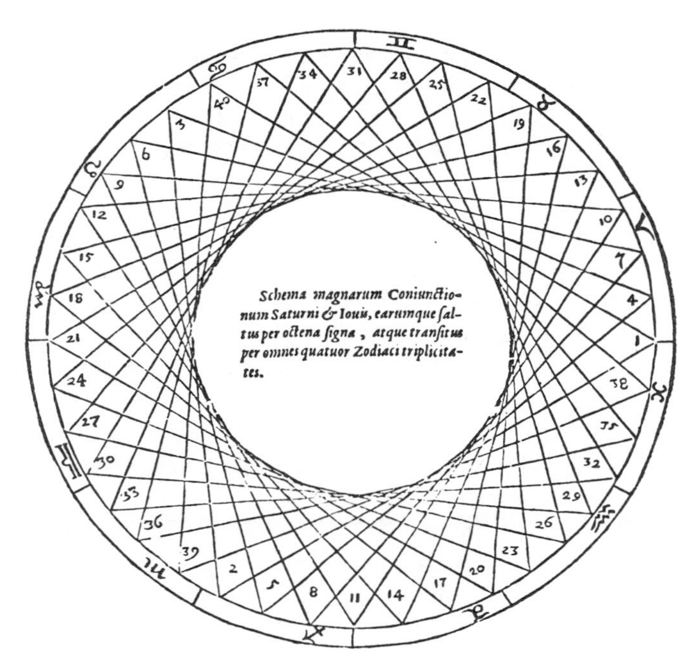


**Kepler's Diagram of the Conjunctions of Jupiter and Saturn**

The number 1 designates the conjunction of the year 1583, 2 that of 1603, 3 that of 1623, 4 that of 1643, etc. Note that these points form a slowly rotating quasi-triangle; this can be seen by drawing a line from 1 to 2 to 3 to 4 to 5, etc. In this figure Kepler noticed that the radius of the inner circle as compared to that of the outer circle is almost exactly in the ratio of the radius of Jupiter's orbit to that of Saturn. This led him to try to discover correlations between the planetary radii and various geometrical figures. He finally hit upon what seemed to him to be a remarkable correlation that explained to his satisfaction both why only six planets exist and why they are positioned at the distances from the sun that they are. In the book he soon published announcing this result, he included a statement he had composed at the time of his discovery:

> The earth is the circle which is the measure of all. Construct a dodecahedron round it. The circle surrounding that will be Mars, round Mars construct a tetrahedron. The circle surrounding that will be Jupiter. Round Jupiter construct a cube. The circle surrounding that will be Saturn. Now construct an icosahedron inside the earth. The circle inscribed within that will be Venus. Inside Venus inscribe an octahedron. The circle inscribed within that will be Mercury.[2]

Kepler presented his "discovery" in his *Mysterium cosmographicum* (1596), which contains the next diagram.

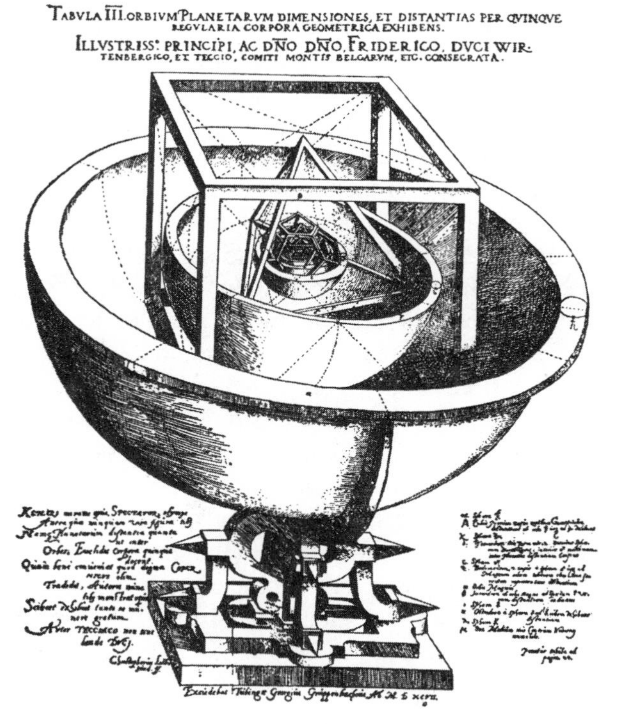


**Kepler's Diagram of the Relation between the Planetary Orbits and the Five Regular Solids**

[2]Johannes Kepler, *Mysterium Cosmographicum. The Secret of the Universe*, trans. A. M. Duncan (New York: Abaris Books, 1981), p. 69.

In constructing this diagram, Kepler set the thickness of each sphere equal to the eccentricities of its planet's orbit. Days and nights went into the calculations checking this correlation. The results were quite good. Except for the case of Saturn, the figures he attained correspond within 5% of the planetary radii as given by Copernicus. This book, which he soon sent off to the aging Danish astronomer **Brahe** and to a young Italian scientist named **Galileo**, is noteworthy not only for Kepler's theory that the solar system was designed on the basis of the **five regular solids**, but also because it contains **one of the earliest endorsements of the Copernican system**.

Let us examine chapter one of that book because in it Kepler explained why he had become convinced of the Copernican system. First of all, he had concluded that Scripture presents no problems for Copernicanism. He planned at one point in time to include his reasoning on this issue, but on the advice of friends jettisoned it, writing to Maestlin: "The whole of astronomy is not worth so much that one of the little ones who follow Christ should be angered." On a philosophical level, he wrote this chapter as a realist advocate of Copernicanism who specifically rejected the instrumentalist or fictionalist approach to astronomy.

The first of Kepler's pro-Copernican astronomical arguments is its ability to "give the reasons for the number, extent, and time of the retrogressions, and why they agree precisely . . . with the position and mean motion of the Sun." Allied to this is his argument that by putting the earth into an orbit, Copernicus made it possible to dispense with the main eccentric circles (their deferents) of the sun, Mercury, and Venus. The main epicycle in the system of Mars as well as that for Jupiter and for Saturn also become unnecessary, Kepler pointed out, in the Copernican system. Moreover, an understanding of the heliocentric theory reveals why the epicycle in Ptolemy's system for Mars must be so large whereas the main Saturnian epicycle is quite small. Bounded elongation is yet another feature explained in the Copernican, but not in the Ptolemaic system. Putting the earth in rotation, Kepler also stressed, makes more sense than having the vast starry vault rotate. In addition, Kepler advanced other arguments of a comparable nature, some more technical, in advocacy of the Copernican system.

Kepler's *Mysterium cosmographicum*, whatever the merits of its hypotheses and whatever reaction readers may have had to its Pythagorean-Platonic approaches, acted on its author to deepen his commitment to astronomy. He decided not to return to Tübingen and thereby forsook his plan to become a Lutheran minister. Counterreformation pressures in dominantly Catholic Graz created tensions, which Kepler escaped in 1600 by joining Brahe's staff in Prague. Before Brahe died in 1601, he set Kepler to work on the problem of accounting for the motions of Mars. Kepler continued to pursue his "War with Mars" during much of the first decade of the seventeenth century, this being possible because he was chosen to succeed Brahe as Imperial Mathematician although at only one-sixth of Brahe's salary. Pausing only to publish in 1604 a major optical treatise, his *Astronomiae pars optica* and some other minor works, Kepler prepared his researches on Mars for publication in a 1609 book.

### *Kepler's "War with Mars"*

In 1609, Kepler published his *Astronomia nova*, the full title of which, if translated into English, is: *A New Astronomy Based on Causation or a Celestial Physics Derived from Investigations of the Motions of Mars Founded on the Observations of the Noble Tycho Brahe.* In a sense it is a history of his "War with Mars," for he recounted in it all his attempts over a number of years to find a mathematical system that would account for the complex motions of Mars. Involved in this search were considerations not only of the mathematics of its motions, but also of its physics. This is to say that, as the title of his book indicates, Kepler was seeking physical reasons for the planet's motions. In particular, he explored the idea that some force may extend out from the sun and move the planets in their orbits.

It would be fascinating to follow, skirmish by skirmish, Kepler's war with Mars, to see how at crucial points Brahe's observations led him to reject theories for which he had high hopes and on which he had spent days calculating, to witness his struggles with epicycles, equants, ovals, ovoids, and ellipses. But only two major results that he published in his *Astronomia nova* need concern us; these we shall call "Kepler's Conjectures."

### *Kepler's "First Conjecture"*

Kepler's first conjecture (which he actually arrived at second) is that **Mars moves in an ellipse with the sun at one focus of the ellipse.** An elementary knowledge of the geometry of ellipses contributes to an understanding of this conjecture. By definition, an ellipse is the locus of points such that the sum of the distances of those points to two fixed points $F_1$ and $F_2$ (the foci of the ellipse) is constant.

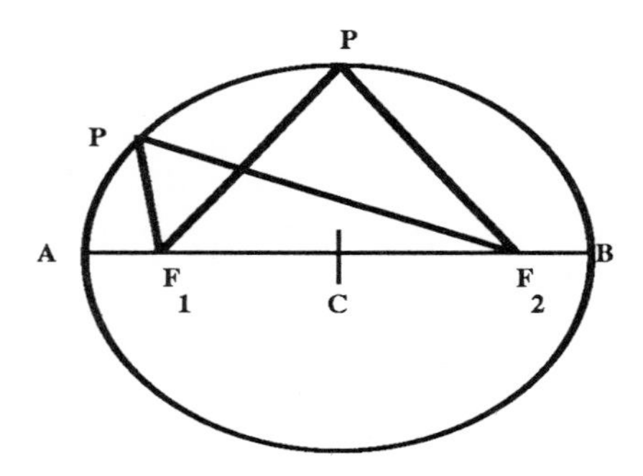


As is evident from the figure, for any point P on the ellipse, the sum of the lengths $F_1P + F_2P$ is constant. Consequently, an ellipse can be constructed by fixing the ends of a string to points $F_1$ and $F_2$ and moving a pencil at the end of the loop thus formed. The **major axis** of the ellipse is the line AB, the line extending through the foci to the opposite sides of the ellipse. The **minor axis** of the ellipse is the line through its center and perpendicular to its major axis. The **eccentricity** (e) of an ellipse is the ratio of the distance $F_1C$ to CA, which is also equal to $F_1F_2/AB$. The eccentricity ranges between 0, in which case the ellipse becomes a circle, and 1, in which case the figure reduces to a straight line. Kepler's first conjecture is, then, that Mars moves in an orbit of elliptical shape with the sun at one focus and nothing at the other focus. As should be evident, Kepler's proposal of an elliptical orbit represented a major break from the tradition of explaining planetary motions by combinations of circular motions.

*Kepler's "Second Conjecture"*

Kepler went on to claim that Mars moves on its ellipse in such a way that a line from the sun to Mars **sweeps out equal areas in equal times.**

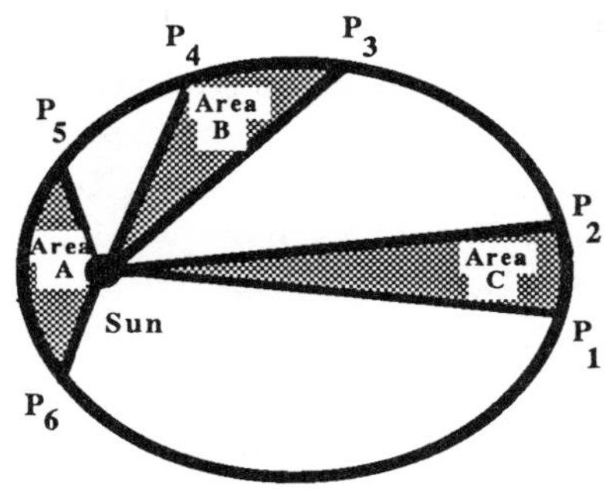


In the figure provided, if areas A, B, and C are equal, then the time taken by Mars in moving from $P_1$ to $P_2$ is equal to the time taken in moving from $P_3$ to $P_4$ or from $P_5$ to $P_6$. You should ask yourself where according to this theory Mars should move most rapidly, where slowest.

Kepler did not refer to these two theories of Mars's motion as conjectures nor are they called by this name in modern astronomy. They are rather known as Kepler's first and second laws of planetary motion. In the years immediately after the 1609 publication of Kepler's book, however, they received a somewhat mixed reception. In fact, they were treated as conjectures even though Kepler marshaled substantial evidence for their correctness, at least as applied to Mars, which was the only planet that he treated in his 1609 volume. A decade later in his *Epitome of Copernican Astronomy* (3 vols., 1617–1621) he did attempt to show their applicability to the other planets.

*Kepler's Later Years*

In 1611, Kepler published a second treatise on optics, his *Dioptrice*, which is notable because in it Kepler worked out the principles of what has come to be known as the refracting telescope. In 1612, he left Prague, becoming district mathematician for Linz. In 1619, he published his *Harmonices mundi* (*Harmonies of the World*) in which he presented an array of harmonies or aesthetically pleasing relationships that he had detected in the universe. We need concern ourselves with only one of these, which is that if the time (T) taken by a planet in moving through one complete orbital cycle is squared, and the distance (D) of the planet from the sun is cubed, then for any planet P, $T_P^2/D_P^3 = C$, where C is a constant that is identical for every planet. It is interesting to test the correctness of this relationship, which later became known as Kepler's "third law," by checking it against the data given below.

| Planet | Sidereal Period (in years) | Distance from Sun (normed on the earth as 1) |
|---|---|---|
| Mercury | .241 | .387 |
| Venus | .615 | .732 |
| Earth | 1.000 | 1.000 |
| Mars | 1.881 | 1.524 |
| Jupiter | 11.862 | 5.203 |
| Saturn | 29.457 | 9.539 |

In 1627, Kepler, drawing on Tycho Brahe's precise planetary observations and his own theories of planetary motion, published his *Tabulae Rudolphinae (Rudolphine Tables),* in which he presented tables from which the position of any planet at any time—past, present, or future—could be calculated. Astronomers came to see that these tables were the most accurate tables of planetary motion devised up to that time. Because he died in 1630, Kepler did not live to see the acclaim his tables attained. He had written his own epitaph:

> I used to measure the skies,
> Now I shall measure the shadows of the earth.
> Skybound was the mind,
> Earthbound the body resides.

# Chapter Nine

## *Galileo Galilei*

**Galileo Galilei**

*Galileo Chronology*

1564 Birth of Galileo Galilei in Pisa on February 15, the son of a cloth merchant and musician, Vincenzio Galilei.

1575 Begins a period of study lasting until 1581 at the monastery of Santa Maria at Vallombrosa; for a time becomes a novice in the Vallombrosan order.

1581 Enrolls at the University of Pisa, initially for medical studies, but soon turns to mathematics.

1589 Having by this time composed a treatise on the center of gravity in bodies and invented a hydrostatic balance, he becomes mathematics professor at Pisa.

1592 Becomes professor of mathematics at the University of Padua.

1597 In a letter to Kepler, Galileo states that he had accepted Copernicanism a few years earlier.

1609 Hearing of a device constructed by a Flemish optician (Hans Lipperhey) by which distant objects can be seen as if nearby, Galileo makes his own telescope and begins using it for astronomical observation.

1610 Publishes his *Sidereus Nuncius (Starry Messenger)* in which he reveals his discovery of the mountainous surface of the moon, the four moons of Jupiter, and the vast number of stars seen by means of the telescope. Resigns from Padua University; moves to Florence with Grand Duke Cosimo II de Medici as patron.

1613 Publishes *Letters on Sunspots* in which he announces his discovery of the phases of Venus, the composite nature of Saturn, and sunspots. By this time, his Copernicanism is beginning to cause controversy.

1615 Cardinal Robert Bellarmine writes to Galileo, urging him to present Copernicanism only in a hypothetical manner.

1616 Holy Office bans Copernicus's book until certain passages are corrected; Galileo warned not to defend Copernicanism.

1620 Congregation of the Index specifies corrections that, if observed, allow reading of Copernicus's book.

1621 Pope Paul V, Cardinal Bellarmine, and Cosimo de Medici all die.

1624 Galileo goes to Rome in hopes of persuading the new pope, Urban VIII, to allow him to write on the Copernican system. Pope agrees, provided Galileo treats Copernicanism hypothetically.

1632 Galileo publishes his *Dialogo . . . sopra i due massimi sistemi del mondo: Tolemaico, e Copernicano (Dialogue on the Two Chief World Systems, Ptolemaic and Copernican)*, vigorously advocating the Copernican system.

1633 Trial of Galileo in which his *Dialogue* is banned; Galileo abjures Copernican system.

1638 Galileo publishes his *Discorsi e dimostrazioni matematiche intorno a due nuove scienze (Discourses on the Two New Sciences)*, which is the beginning of modern mechanics.

1642 Galileo dies on January 8.

*Galileo and Early Telescopic Astronomy*

Galileo contributed to science in many ways. This chapter focuses on his work during the decade immediately following the invention of the telescope in 1608. It should, however, be kept in mind that some of Galileo's most important discoveries were first presented in the 1630s, a period beyond the scope of the present book.

Although Galileo did not invent the telescope, he, more than anyone else in the years immediately after its invention, exploited its potential as a method of obtaining astronomical information. Galileo's telescopes consisted of a concave eyepiece joined to a convex objective lens. Such telescopes are called **Galilean** telescopes to distinguish them from telescopes that have a convex lens for their eyepiece. The theory of the latter type of telescope was first worked out by **Kepler** around 1610, but such telescopes, now called Keplerian or astronomical telescopes, were first built only toward the end of the second decade of the seventeenth century. Diagrammatically, the two types are as follows:

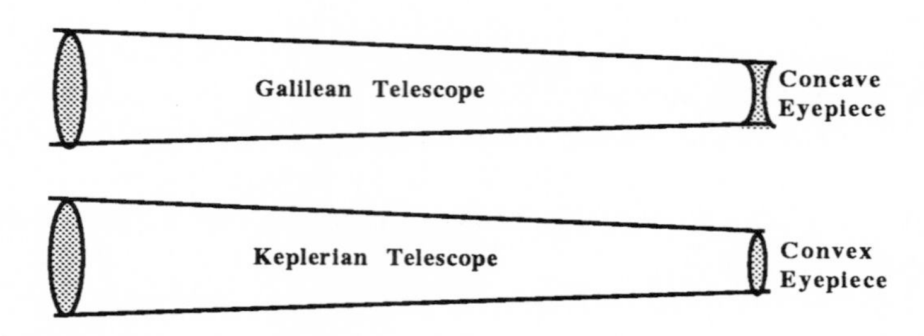


Opera glasses and very low cost telescopes are usually of Galilean design. This is appropriate because this type of lens system gives an erect image, but does not permit high quality observation. Keplerian telescopes give an inverted image, inconvenient at an opera or a football game, but presenting no problem in astronomical observation. Moreover, Keplerian telescopes can be constructed so as to permit the resolution of fine detail, e.g., the surface features of the moon.

These considerations are relevant because it is important not to overestimate the quality of Galileo's early telescopes, as would be easy to do from Galileo's statements in his *Starry Messenger*, which he published in 1610 to report on the discoveries he had made since mid-1609 when he first began observing with his telescope. For example, he stated in that booklet that his instrument made the moon's diameter appear "almost thirty times larger, its surface nearly nine hundred times, and its volume twenty-seven thousand times as large as when viewed with the naked eye."[1] Put in modern terms, the magnification of Galileo's telescope was 30. The larger figures cited by him derived from the

---

[1]This and the other quotations in this chapter from Galileo's *Starry Messenger* are quoted with permission from the translation of that work published in *Discoveries and Opinions of Galileo*, translated with an introduction and notes by Stillman Drake (New York: Doubleday Anchor, 1957), pp. 20–58.

facts that the area of a circle is proportional to its diameter squared and the volume of a sphere is proportional to its diameter cubed. Moreover, the objective lenses of his instruments were no larger than about 2.5 inches in diameter, which seriously limited their power to resolve fine detail. Given these limitations, it is in some ways surprising that Galileo achieved as much as he did with his telescopes.

### *Galileo's Telescopic Discoveries as of 1610*

The observational results Galileo reported in his *Starry Messenger* of 1610 were startlingly new and shocking to many of his contemporaries. In considering them, it is best to use a measure of historical imagination, necessary in these days when children of five have seen pictures of the moon's surface taken within a few feet of it and have been transported by television to within a matter of miles from Jupiter's moons. It is also important to ask: to what extent, if at all, did Galileo's observations help the Copernican cause? Could they have been incorporated into the geocentric system?

Regarding the **moon**, Galileo reported that observations of its dark and light regions led him to the conviction that

> . . . the surface of the moon is not smooth, uniform, and precisely spherical as a great number of philosophers believe it (and the other heavenly bodies) to be, but is uneven, rough, and full of cavities and prominences, being not unlike the face of earth, relieved by chains of mountains and deep valleys.

With a measure of caution, he commented that "if anyone wished to revive the old Pythagorean opinion that the moon is like another earth, its brighter parts might very fitly represent the surface of the land and its darker region that of the water."

Galileo not only reported detection of numerous mountains on the moon, but also attempted to measure the heights of some peaks. He explained his method in the following statement:

> I had often observed, for various situations of the moon with respect to the sun, that some summits within the shadowy portion appeared lighted, though lying some distance

from the boundary of the light. By comparing this separation to the whole diameter of the moon, I found that it sometimes exceeded one-twentieth of the diameter. Accordingly, let CAF be a great circle of the lunar body, E its center, and CF a diameter, which is to the diameter of the earth as two is to seven.

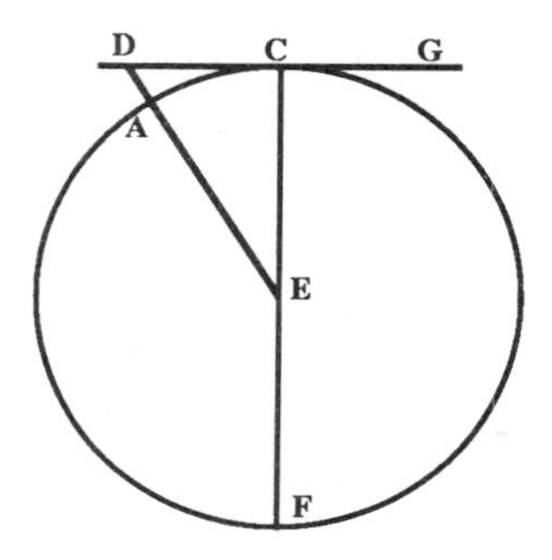


Since according to very precise observations the diameter of the earth is seven thousand miles, CF will be two thousand, CE one thousand, and one-twentieth of CF will be one hundred miles. Now let CF be the diameter of the great circle which divides the light part of the moon from the dark part (for because of the very great distance of the sun from the moon, this does not differ appreciably from a great circle), and let A be distant from C by one-twentieth of this. Draw the radius EA, which, when produced, cuts the tangent line GCD (representing the illuminating ray) to the point D. Then the arc CA, or rather the straight line CD, will consist of one hundred units whereof CE contains one thousand, and the sum of the squares of DC and CE will be 1,010,000. This is equal to the square of DE; hence ED will exceed 1,004, and AD will be more than four of those units of which CE contains one thousand. Therefore the altitude AD on the moon, which represents a summit reaching up to the solar ray GCD and standing at the distance CD from C, exceeds four miles. But on the earth we have no mountains which reach to a perpendicular height of even one mile. Hence it is quite clear that the prominences on the moon are loftier than those on the earth.

It is interesting to note that Galileo's claim that some lunar mountains are four times higher than any terrestrial mountain is incorrect and for a curious reason. His estimate that lunar mountains are as high as four Italian miles is not seriously off; lunar mountains according to current data do attain heights of about 25,000 feet. Nonetheless, Galileo was seriously in error as to the maximum height of terrestrial peaks; Mount Everest in Tibet at 29,028 feet is probably taller than any lunar mountain.

Another feature of Galileo's *Starry Messenger* was his inclusion of four drawings of the moon's surface.

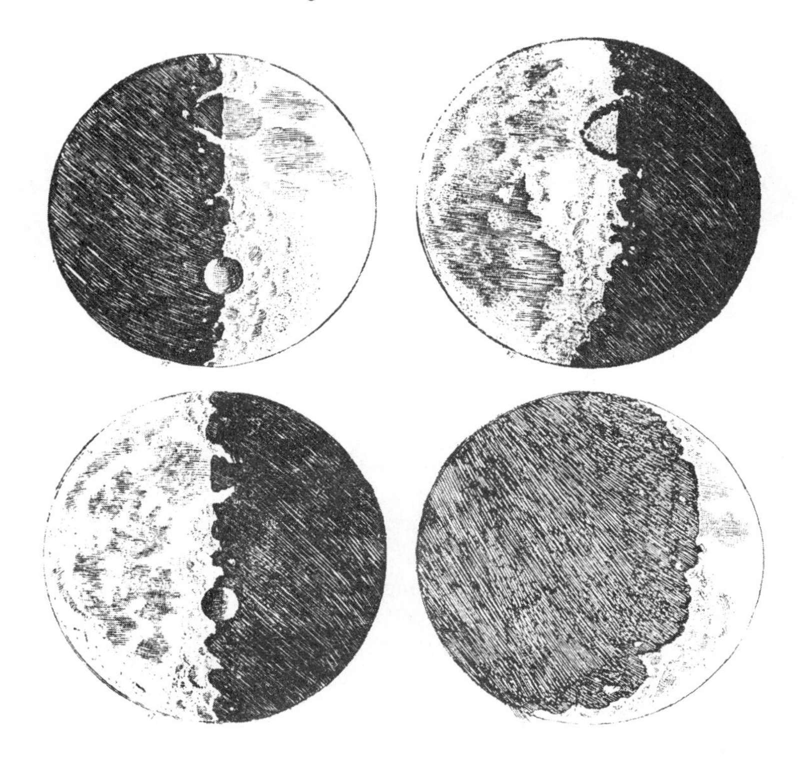

**Galileo's Moon Drawings from His *Starry Messenger***

The question of the accuracy of these drawings is much disputed; for example, Arthur Koestler in his book *The Sleepwalkers* (1959) charged in regard to these drawings that "the huge dark spot under the moon's equator, surrounded by mountains . . . simply does not exist." The astronomer Zdenek Kopal in his *Mapping the Moon* (1974) asserted concerning Galileo's lunar observations that "none of the features recorded on this (and other) drawings of the Moon can be safely identified with any known markings of the lunar landscape." Historians of science, astronomers, and philosophers of science have now produced a large literature on this topic, some supporting Galileo, others agreeing with Koestler and Kopal. This controversy is relevant to the question of the degree of objectivity that can be attained in astronomical observation.

Regarding **stars**, Galileo's telescope revealed a rich array of new observations. He first of all reported that the telescope does not magnify stars in the same way that it increases the apparent diameter of, for instance, the moon. Although stars appear much brighter in a telescope, their diameters do not proportionally increase. Comparing the telescopic view of stars with planets, he stated:

> The planets show their globes perfectly round and definitely bounded, looking like little moons, spherical and flooded all over with light; the fixed stars are never seen to be bounded by a circular periphery, but have rather the aspect of blazes whose rays vibrate about them and scintillate a great deal.

Even more striking is the fact that Galileo saw vast numbers of stars never seen previously. When examining the constellation Orion, for example, Galileo found:

> There are more than five hundred new stars distributed among the old ones within limits of one or two degrees of arc. Hence to the three stars in the belt of Orion and the six in the Sword which were previously known, I have added eighty adjacent stars. . . .

Directing his telescope to the Milky Way, he found it consists of

> . . . nothing but congeries of innumerable stars grouped together in clusters. Upon whatever part of it the telescope is directed, a vast crowd of stars is immediately present to view. Many of them are rather large and quite bright, while the number of smaller ones is quite beyond calculation.

Galileo devoted the final portion of his *Starry Messenger* to recounting what he described as his "most important" discovery: "the disclosure of four PLANETS never seen from the creation of the world up to our own time. . . ." These were the **four moons of Jupiter**, which, in honor of his patron, he named the "four Medicean planets." Nothing can more effectively convey the excitement generated by this discovery than an extended quotation from Galileo's discovery report:

> On the seventh day of January in this present year 1610, at the first hour of night, when I was viewing the heavenly bodies with a telescope, Jupiter presented itself to me; and because I had prepared a very excellent instrument for myself, I perceived (as I had not before, on account of the weakness of my previous instrument) that beside the planet there were three starlets, small indeed, but very bright. Though I believed them to be among the host of fixed stars, they aroused my curiosity somewhat by appearing to lie in an exact straight line parallel to the ecliptic, and by their being more splendid than others of their size. Their arrangement with respect to Jupiter and each other was the following:
>
> *East* * * O * *West*
>
> that is, there were two stars on the eastern side and one to the west. The most easterly star and the western one appeared larger than the other. I paid no attention to the distances between them and Jupiter, for at the outset I thought them to be fixed stars, as I have said. But returning to the same investigation on January eight—led by what, I do not know—I found a very different arrangement. The three starlets were now all to the west of Jupiter, closer together, and at equal intervals from one another as shown in the following sketch:
>
> *East* O * * * *West*
>
> At this time, though I did not yet turn my attention to the way the stars had come together, I began to concern myself with the question how Jupiter could be east of all these stars when on the previous day it had been west of two of them. I commenced to wonder whether Jupiter was not moving eastward at that time, contrary to the computations of the astronomers, and had got in front of them by that motion. Hence it was with great interest that I awaited the next night.

But I was disappointed in my hopes, for the sky was then covered with clouds everywhere.

On the tenth of January, however, the stars appeared in this position with respect to Jupiter:

*East* * * O *West*

that is, there were but two of them, both easterly, the third (as I supposed) being hidden behind Jupiter. As at first, they were in the same straight line with Jupiter and were arranged precisely in the line of the zodiac. Noticing this, and knowing that there was no way in which such alterations could be attributed to Jupiter's motion, yet being certain that these were still the same stars I had observed (in fact no other was to be found along the line of the zodiac for a long way on either side of Jupiter), my perplexity was now transformed into amazement. I was sure that the apparent changes belonged not to Jupiter but to the observed stars, and I resolved to pursue this investigation with greater care and attention.

And thus, on the eleventh of January, I saw the following disposition:

*East* * * O *West*

There were two stars, both to the east, the central one being three times as far from Jupiter as from the one farther east. The latter star was nearly double the size of the former, whereas on the night before they had appeared approximately equal.

I had now decided beyond all question that there existed in the heavens three stars wandering about Jupiter as do Venus and Mercury about the sun, and this became plainer than daylight from observation on similar occasions which followed. Nor were there just three such stars; four wanderers complete their revolutions about Jupiter, and of their alterations as observed more precisely later on we shall give a description here. Also I measured the distances between them by means of the telescope, using the method explained before. Moreover I recorded the times of the observations, especially when more than one was made during the same night—for the revolutions of these planets are so speedily completed that it is usually possible to take even their hourly variations.

Thus on the twelfth of January at the first hour of night I saw the stars arranged in this way:

*East* * * O * *West*

The most easterly star was larger than the western one, though both were easily visible and quite bright. Each was about two minutes of arc distant from Jupiter. The third star was invisible at first, but commenced to appear after two hours; it almost touched Jupiter on the east, and was quite small. All were on the same straight line directed along the ecliptic.

On the thirteenth of January four stars were seen by me for the first time, in this situation relative to Jupiter:

*East* * O * * * *West*

Three were westerly and one was to the east; they formed a straight line except that the middle western star departed slightly toward the north. The eastern star was two minutes of arc away from Jupiter, and the intervals of the rest from one another and from Jupiter were about one minute. All the stars appeared to be of the same magnitude, and though small were very bright, much brighter than the fixed stars of the same size.

Galileo continued his observations until March 2, by which time he had obtained sixty-four sightings of the moons of Jupiter.

Galileo did not hesitate to suggest the importance of his discovery for the Copernican debate although it was only in 1613 that he first publicly endorsed that system. He did, however, state in his 1610 work that in the four Jupiterian satellites,

> . . . we have a fine and elegant argument for quieting the doubts of those who, while accepting with tranquil mind the revolutions of the planets about the sun in the Copernican system, are mightily disturbed to have the moon alone revolve about the earth and accompany it in an annual rotation about the sun. Some have believed that this structure of the universe should be rejected as impossible. But now we have not just one planet rotating about another while both run through a great circle around the sun; our own eyes show us four stars which wander around Jupiter as does the moon around the earth, while all together trace out a grand revolution about the sun in the space of twelve years.

Not surprisingly, Galileo's *Starry Messenger* created a sensation.

*Galileo's Other Pre-1615 Telescopic Discoveries*

It is not infrequently stated that between late 1610 and 1613, Galileo discovered the phases of Venus, the rings of Saturn, and sunspots. Let us examine each of these claims.

**1. Phases of Venus**

On December 11, 1610, Galileo sent a letter containing an anagram to one of the Medici. Seventeenth-century scientists sometimes used anagrams to establish priority for a discovery without revealing the discovery itself. The anagram Galileo used was "Haec immatura a me iam frustra leguntur oy," which can be translated as "These are at present too young to be read by me oy." Galileo's secret message in the anagram was: "Cynthiae figuras aemulatur Mater Amorum." Translated this reads: "Cynthia's figures are imitated by the Mother of Love." This deciphered means that Venus (the Mother of Love) has phases like the moon (Cynthia). Galileo was the first to make this discovery, which he included in his 1613 *Letters on Sunspots*. In particular, Galileo stated that Venus is sometimes seen as small and bright over its entire surface, sometimes as larger and in a half phase, and sometimes still larger but crescent shaped. He took this to be proof that Venus orbits the sun. As he wrote: "With absolute necessity we shall conclude, in agreement with the theories of the Pythagoreans and of Copernicus, that Venus revolves about the sun just as do all the planets."

**2. Rings of Saturn**

In summer, 1610, Galileo made a further discovery, which he communicated to possibly the most enthusiastic of all readers of his *Starry Messenger*. This was Johannes Kepler, who learned: "SMAISMRMILMEPOETALEUMIBUNENUGTTAUIRAS." Kepler puzzled over this anagram, finally finding in it "Salve umbistineum geminatum Martia proles," which fits the anagram and which can be translated as: "Hail, burning twin, offspring of Mars!" Alas, Kepler was wrong; Galileo had not found the two moons for Mars, which were discovered only in 1877. Galileo finally revealed what he had discovered: "Altissimum planetam tergeminum observavi" or "I have seen the uppermost planet triple." In his *Letters on Sunspots* (1613), Galileo set out his Saturnian observations in opposition to those reported by the

Jesuit Christopher Scheiner, who is referred to as "Apelles" in the following passage:

> For the same reason I have resolved not to put anything around Saturn except what I have already observed and revealed—that is, two small stars which touch it, . . . and in which no alteration has ever yet been seen to take place and in which none is to be expected in the future, barring some very strange event remote from every other motion known to or even imagined by us. But as to the supposition of Apelles that Saturn is sometimes oblong and sometimes accompanied by two stars on its flanks, Your Excellency may rest assured that this results either from the imperfection of the telescope or the eye of the observer, for the shape of Saturn is thus:
>
> 
>
> as shown by perfect vision and perfect instruments, but appears thus:
>
> 
>
> where perfection is lacking, the shape and distinction of the three stars being imperfectly seen. I, who have observed it a thousand times at different periods with an excellent instrument, can assure you that no change whatever is to be seen in it.[2]

Having expressed himself so strongly, Galileo was much taken aback when in very late 1612 the Saturnian appendages vanished! In a letter of December 1, 1612, Galileo conveyed his amazement and distress at their disappearance.

> Now what can be said of this strange metamorphosis? That the two lesser stars have been consumed, in the manner of the sunspots? Has Saturn devoured his children? Or was it indeed an illusion and a fraud with which the lenses of my telescope deceived me for so long—and not only me, but many others

[2]This quotation and the next are given with permission from Galileo's *Letters on Sunspots* as published in *Discoveries and Opinions of Galileo*, translated with an introduction and notes by Stillman Drake (New York: Doubleday Anchor, 1957); see pp. 101–2; 143–4.

> who have observed it with me? Perhaps the day has arrived when languishing hope may be revived in those who, led by the most profound reflections, once plumbed the fallacies of all my new observations and found them to be incapable of existing!

The story of the problems Saturn created for early telescopic astronomers could be carried much further, but this is unnecessary. What was happening was that Galileo in 1610 had seen the ring of Saturn, but by late 1612, the plane of the ring as seen from the earth had tilted to the point that Galileo's line of sight corresponded with the plane of the ring, making it invisible!

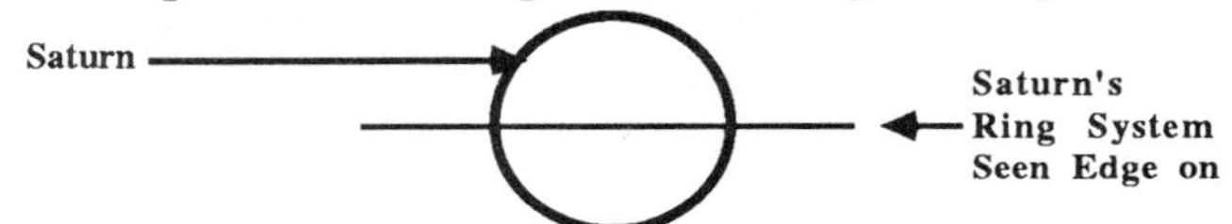


In time, the ring reappeared, but it was only in 1659, when Christiaan Huygens published his *Systema Saturnium,* that this mystery was cleared up. In that book, Huygens presented a diagram that explained what had been happening.

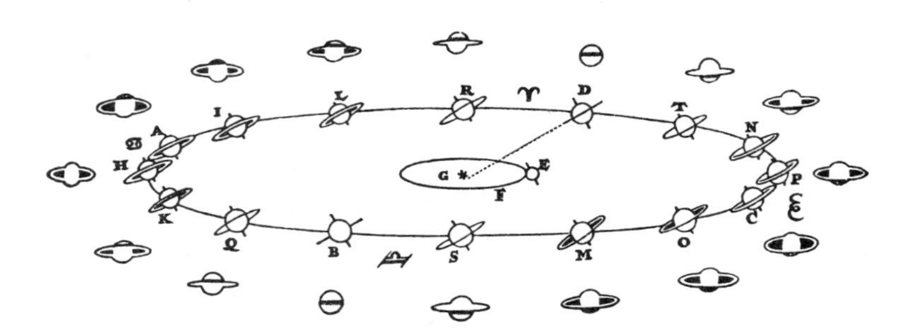

**Huygens's Diagram of the Saturnian System**

Huygens also included in his volume a composite representation of thirteen drawings of Saturn made by astronomers from 1610 to 1650 (see illustration). A glance at this figure suggests the uncertainties that beset early telescopic observation. The astronomers and the year(s) in which each made the drawings are as follows: I. Galileo (1610); II. Scheiner (1614); III. Riccioli (1641 or 1643); IV–VII. Hevelius (based partly on theory); VIII–

IX. Riccioli (1646–1650); X. Divini (1646); XI. Fontana (1636); XII. Gassendi (1646); XIII. Fontana (1644–1645).

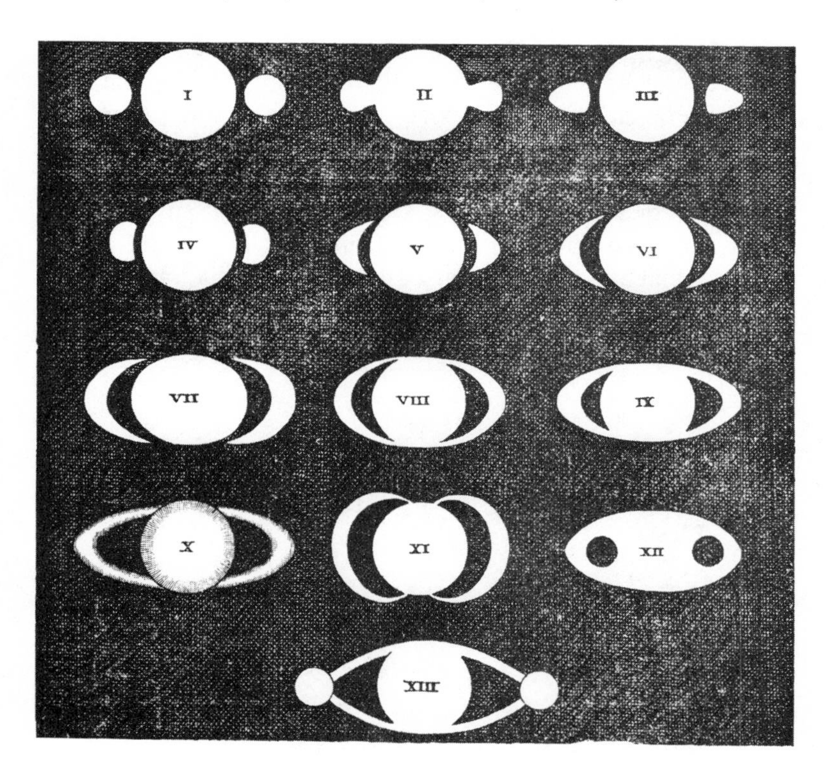


**Huygens's Composite Representation of Views of Saturn**

## 3. Sunspots

Three astronomers are generally regarded as rivals for the title to the discovery of sunspots. They are Galileo, Scheiner, and David Fabricius. All had seen them by 1612, Fabricius being the first to publish his discovery. To carry the question of priority beyond these statements would necessitate entering into historical complexities that need not now be considered. Allied to this dispute was the question of whether sunspots occur on the surface of the sun or are due to objects, small planets, for example, orbiting above it. And allied to that question was a controversy over whether the sun can change.

## *Conclusion*

The sensation created by Galileo's *Starry Messenger* extended not only to scientists, but also to the educated public as well. Evidences of its influence are to be found in, for example, the English, French, and Italian poetry from that period. A crucial question to be asked about Galileo's discoveries is whether or not they in any way proved the Copernican system. You should ponder this question, attempting to decide which, if any, had that effect and if so, precisely how. This leads to the still larger question that is implicitly raised by the next diagram, which appeared in 1660 in Athanasius Kircher's *Iter extaticum coeleste* .

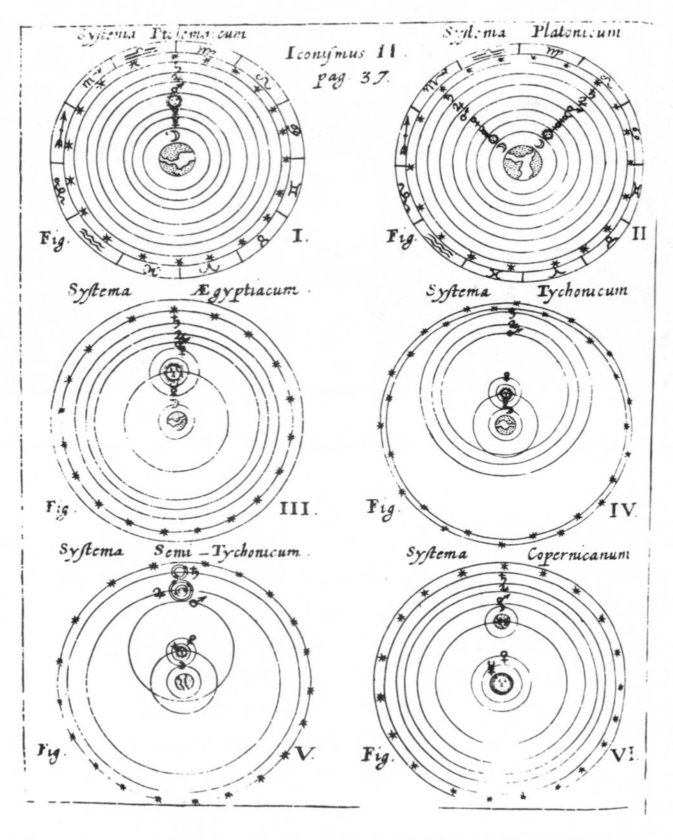


**Kircher's Diagram of Six Astronomical Systems**

The two uppermost systems, those Kircher referred to as the Ptolemaic and Platonic systems, differ from each other only in that the former places the sun's orbit beyond the orbits of Mercury and Venus, whereas in the second the sun's orbit is placed inside the orbits of those two planets. In Kircher's Egyptian system, the orbits of Mercury and Venus are centered on the sun, the orbit of which is centered on a possibly rotating earth. In Kircher's Tychonic and semi-Tychonic systems, the centers of the orbits of Jupiter and Saturn are positioned differently; in the former, the orbits of those two planets are centered on the sun, whereas in the latter, they are centered on the earth, which is viewed as stationary. In the Copernican system (bottom right), the earth not only rotates but is itself placed in an orbit centered on the sun. Kircher's diagram not only points to the problem of the relativity of motion, but also raises the important question central to this book: **given the evidence available in 1615, which system of the world had at that time the strongest claim for acceptance?**

# Epilogue

## *Some Quotations Concerning Astronomy in the Copernican Revolution Period*

*Sixteenth Century*

Martin Luther (1483–1546)

From his *Tischreden (Tabletalk)*

There was mention of a certain new astrologer who wanted to prove that the earth moves and not the sky, the sun, and the moon. This would be as if somebody were riding on a cart or in a ship and imagined that he was standing still while the earth and the trees were moving. [Luther remarked:] "So it goes now. Whoever wants to be clever must agree with nothing that others esteem. He must do something of his own. This is what that fellow does who wishes to turn the whole of astronomy upside down. Even in these things that are thrown into disorder I believe the Holy Scriptures, for Joshua commanded the sun to stand still and not the earth."[1]

Philip Melanchthon (1497–1560)

From his *Initia doctrinae physicae* (1550)

. . . the Son of God is One; our master Jesus Christ was born, died, and resurrected in this world. Nor does He manifest Himself elsewhere, nor elsewhere has He died or resurrected. Therefore it must not be imagined that Christ died and was resurrected more often, nor must it be thought that in any other world without the knowledge of the Son of God, that men would be restored to eternal life.[2]

---

[1]As translated in Edward Rosen, *Copernicus and the Scientific Revolution* (Malabar, FL, 1984), pp. 182–3.

[2]As quoted in Steven J. Dick, *Plurality of Worlds: The Origins of the Extraterrestrial Life Debate from Democritus to Kant* (Cambridge, England, 1982), p. 89.

## Michel de Montaigne (1533–1592)

From *Apologie de Raymond Sebond* (1580)
Copernicus has so well established [his] doctrine that he uses it . . . for all astronomical computations. What shall we draw from that, if not that we should not worry which of the two systems is true? Who knows but that a third opinion . . . may not overthrow the two previous ones?[3]

## Giordano Bruno (1548–1600)

From his *La cena de le ceneri (The Ash Wednesday Supper)* (1584)
[Copernicus] in regard to natural judgment was far superior to Ptolemy, Hipparchus, Eudoxus, and all the others who walked in the footsteps of these; a man who had to liberate himself from some false presuppositions of the common and commonly accepted philosophy, or perhaps I should say blindness. But for all that he did not move too much beyond them; being more intent on the study of mathematics than of nature, he was not able to go deep enough and penetrate beyond the point of removing from the way the stumps of inconvenient and vain principles, so as to resolve completely the difficult objections, and to free both himself and others from so many vain investigations and to set attention firmly on things constant and certain.[4]

From his *De l'infinito universo et mondi (On the Infinite Universe and Worlds)* (1584)
Thus is the excellence of God magnified and the greatness of his kingdom made manifest; he is glorified not in one, but in countless suns; not in a single earth, a single world, but in a thousand thousand, I say in an infinity of worlds.[5]

---

[3]Michel de Montaigne, *In Defense of Raymond Sebond*, trans. A. H. Beattie (New York, 1959), p. 86.

[4]As given in Giordano Bruno, *The Ash Wednesday Supper*, trans. with an introduction by Stanley L. Jaki (The Hague, 1975), pp. 56–7.

[5]As given in Dorothea Waley Singer, *Giordano Bruno: His Life and Thought with Annotated Translation of His Work* On the Infinite Universe and Worlds (New York, 1950), p. 246

*Seventeenth Century*

William Shakespeare (1564–1616)

From *The Merchant of Venice* (Printed in 1600)

Look, how the floor of heaven
Is thick inlaid with the patines of bright gold:
There's not the smallest orb which thou behold'st
But in his motion like an angel sings,
Still quiring to the young-eyed cherubins,—
Such harmony is in immortal souls;
But whilst this muddy vesture of decay
Doth grossly close it in, we cannot hear it.—

Act V, Scene 1

From *Troilus and Cressida* (Printed in 1609)

The heavens themselves, the planets, and this centre,
Observe degree, priority, and place,
Insisture, course, proportion, season, form,
Office and custom, in all line of order:
And therefore is the glorious planet Sol
In noble eminence enthroned and sphered
Amidst the other; whose med'cinable eye
Corrects the ill aspects of planets evil,
And posts, like the commandment of a king,
Sans check, to good and bad: but, when the planets,
In evil mixture to disorder wander,
What plagues, and what portents, what mutiny,
What raging of the sea, shaking of earth,
Commotion in the winds, frights, changes, horrors,
Divert and crack, rend and deracinate
The unity and married calm of states
Quite from their fixture! O, when degree is shaked
Which is the ladder to all high designs,
The enterprise is sick!

Act I, Scene 3

Johannes Kepler (1571–1630)

From his 1610 letter to Galileo, whose *Starry Messenger* Kepler had just received

. . . I rejoice that I am to some extent restored to life by your work. If you had discovered any planets revolving around one of the fixed stars, there would now be waiting for me chains and a prison among Bruno's innumerabilities. I should rather say, exile to his infinite space. Therefore, by reporting these four planets revolve, not around one of the fixed stars, but around the planet Jupiter, you have for the present freed me from the great fear which gripped me as soon as I had heard about your book. . . .[6]

John Donne (1571/2–1631)

From "An Anatomie of the World, The First Anniversary" (1611)

And now the Springs and Sommers which we see,
Like sonnes of women after fiftie bee.
And new Philosophy [Science] calls all in doubt,
The Element of fire is quite put out;
The Sun is lost, and th'earth, and no mans wit
Can well direct him, where to looke for it.
And freely men confesse that this world's spent,
When in the Planets, and the Firmament
They seeke so many new; then see that this
Is crumbled out againe to his Atomies.
'Tis all in peeces, all cohaerence gone;
All just supply, and all Relation. . . .

lines 203–14

For the worlds beauty is decai'd, or gone,
Beauty, that's colour, and proportion.
We thinke the heavens enjoy their Sphericall,
Their round proportion embracing all.
But yet their various and perplexed course,
Observ'd in divers ages, doth enforce
Men to finde out so many Eccentrique parts,

[6] As translated in *Kepler's Conversation with Galileo's Sidereal Messenger*, trans. Edward Rosen (New York, 1965), pp. 36–7.

Such divers downe-right lines, such overthwarts,
As disproportion that pure forme: It [the new science] teares
The Firmament in eight and fortie sheires,
And in these Constellations then arise
New starres, and old doe vanish from our eyes:
As though heav'n suffered earthquakes, peace or war,
When new Towers rise, and old demolish't are.
They have impal'd within a Zodiake
The free-borne Sun, and keepe twelve Signes awake
To watch his steps; the Goat and Crab controule,
And fright him backe, who else to either Pole
(Did not these Tropiques fetter him) might runne:
For his course is not round; nor can the Sunne
Perfit a Circle, or maintaine his way
One inch direct; but where he rose to-day
He comes no more, but with a couzening line,
Steales by that point, and so is Serpentine:
And seeming weary with his reeling thus,
He meanes to sleepe, being now falne nearer us.
So, of the Starres which boast that they do runne
In Circle still, none ends where he begun.
All their proportion's lame, it sinkes, it swels.
For of Meridians, and Parallels,
Man hath weav'd out a net, and this net throwne
Upon the Heavens, and now they are his owne.
Loth to goe up the hill, or labour thus
To goe to heaven, we make heaven come to us.
We spur, we reine the starres, and in their race
They're diversly content t'obey our pace.
But keepes the earth her round proportion still?
lines 249–85

From "A Funerall Elegie" (which is part of the same poem)
But, as when heaven lookes on us with new eyes,
Those new starres every Artist exercise,
What place they should assigne to them they doubt,
Argue, and agree not, till those starres goe out. . . .
lines 67–70

From "Of the Progresse of the Soule, The Second Anniversarie" (1612)

. . .Twixt heaven, and earth: she [the soul of the late Elizabeth Drury] stayes not in the ayre,
To looke what Meteors there themselves prepare;
She carries no desire to know, nor sense,
Whether th'ayres middle region be intense;
For th'Element of fire, she doth not know,
Whether she past by such a place or no;
She baits not at the Moone, nor cares to trie
Whether in that new world, men live, and die.
*Venus* retards her not, to'enquire, how shee
Can, (being one starre) *Hesper,* and *Vesper* bee;
Hee that charm'd *Argus* eyes, sweet *Mercury,*
Workes not on her, who now is growne all eye;
Who, if she meet the body of the Sunne,
Goes through, not staying till his course be runne;
Who finds in *Mars* his Campe no corps of Guard;
Nor is by *Jove,* nor his father barr'd;
But ere she can consider how she went,
At once is at, and through the Firmament.
And as these starres were but so many beads
Strung on one string, speed undistinguish'd leads
Her through those Spheares, as through the beads, a string,
Whose quicke succession makes it still one thing. . . .

lines 189–210

From "A Valediction: Forbidding Mourning"

Moving of th'earth brings harmes and feares,
Men reckon what it did and meant. . . .

lines 9–10

George Herbert (1593–1633)

From "Divinitie"

As men, for fear the starres should sleep and nod,
And trip at night, have spheres suppli'd;
As if a starre were duller than a clod,
Which knows his way without a guide. . . .

Then burn thy Epicycles, foolish man;
    Burn all thy spheres, and save thy head.
Faith needs no staffe of flesh, but stoutly can
    To heav'n alone both go, and leade.

From "The Temper," I
Although there were some fourtie heav'ns, or more
    Sometimes I peere above them all;
    Sometimes I hardly reach a score,
        Sometimes to hell I fall.

O rack me not to such a vast extent;
    Those distances belong to thee:
    The world's too little for thy tent,
    A grave too big for me.

William Drummond of Hawthornden (1585–1649)

From "The Cypresse Grove" (1623)
The Earth is found to move, and is no more the centre of the Universe. . . . Some afirme there is another World of men and sensitive Creatures, with cities and palaces in the Moone. . . . Thus Sciences . . . have become Opiniones, nay Errores, and leade the Imagination in a thousand Labyrinthes.

Pierre de Cazre (1589–1664)

From a 1642 letter to Pierre Gassendi
Ponder less on what you yourself perhaps think than on what will be the thoughts of the majority of others who carried away by your authority or your reasons, become persuaded that the terrestrial globe moves among the planets. They will conclude at first that, if the earth is doubtless one of the planets and also has inhabitants, then it is well to believe that inhabitants exist on the other planets and are not lacking in the fixed stars, that they are even of a superior nature and in proportion as the other stars surpass the earth in size and perfection. This will raise doubts about Genesis which says that the earth was made before the stars and that they were created on the fourth day to illuminate the earth

and measure the seasons and years. Then in turn *the entire economy of the Word incarnate and of scriptural truth will be rendered suspect.*[7]

Henry More (1614–1687)

From *Democritus Platonissans, or, An Essay upon the Infinity of Worlds out of Platonick Principles* (1646)
And what is done in this Terrestriall starre
The same is done in every Orb beside.
Stanza 13

[God did his] endlesse overflowing goodness spill
In every place; which streight he did contrive
Int' infinitie severall worlds. . . .
Stanza 50

. . . long ago there Earths have been
Peopled with men and beasts before this Earth,
And after this shall others be again
And other beasts and other humane birth. . . .
Another Adam once received breath
And still another in endless repedation
And this must perish once in finall conflagration.
Stanza 76

René Descartes (1596–1650)

From his June 6, 1647 letter to Chanut
I do not see at all that the mystery of the Incarnation, and all the other advantages that God has brought forth for man obstruct him from having brought forth an infinity of other very great advantages for an infinity of other creatures. And although I do not at all infer from this that there would be intelligent creatures in the stars or elsewhere, I also do not see that there would be any reason by which to prove that there were not; but I always leave

[7]As translated in Michael J. Crowe, *The Extraterrestrial Life Debate 1750–1900: The Idea of a Plurality of Worlds from Kant to Lowell* (Cambridge, England, 1986), pp. 17–18.

undecided questions of this kind rather than denying or affirming anything.[8]

Sir William Davenant (1606–1668)

From *Gondibert* (1651)
Others with Optick Tubes the Moons scant face
(Vaste Tubes, which like long Cedars mounted lie)
Attract through Glasses to so neer a space
As if they came not to survey, but prie.

Nine hasty Centuries are now fulfill'ed,
Since Opticks first were known to *Astragon;*
By whom the Moderns are become so skill'd,
They dream of seeing to the Maker's Throne.

And wisely *Astragon,* thus busy grew,
To seek the Stars remote societies;
And judge the walks of th'old, by finding new;
For Nature's law, in correspondence lies.

Man's pride (grown to Religion) he abates,
By moving our lov'd Earth; which we think fix'd;
Think all of it, and it to none relates;
With others motion scorn to have it mix'd:

As if 'twere great and stately to stand still
Whilst other Orbes dance on; or else think all
Those vaste bright Globes (to shew God's needless skill)
Were made but to attend our little Ball.
Second Book, Canto V, Stanzas 16–20

Blaise Pascal (1623–1662)

From *Pensées* (first published in 1670)
#72 (Brunschvicg numeration) = #199 (Lafuma numeration)
Let man then contemplate the whole of nature in her full and grand

---

[8]As translated in Crowe, *Debate*, p. 16.

majesty, and turn his vision from the low objects which surround him. Let him gaze on that brilliant light, set like an eternal lamp to illumine the universe; let the earth appear to him a point in comparison with the vast circle described by the sun; and let him wonder at the fact that this vast circle is itself but a very fine point in comparison with that described by the stars in their revolution round the firmament. But if our view be arrested there, let our imagination pass beyond; it will sooner exhaust the power of conception than nature that of supplying material for conception. The whole visible world is only an imperceptible atom in the ample bosom of nature. No idea approaches it. We may enlarge our conceptions beyond all imaginable space; we only produce atoms in comparison with the reality of things. It is an infinite sphere, the centre of which is everywhere, the circumference nowhere. In short it is the greatest sensible mark of the almighty power of God, that imagination loses itself in that thought.

Returning to himself, let man consider what he is in comparison with all existence; let him regard himself as lost in this remote corner of nature; and from the little cell in which he finds himself lodged, I mean the universe, let him estimate at their true value the earth, kingdoms, cities, and himself. What is a man in the Infinite?[9]

#205 (Brunschvicg) = #68 (Lafuma)

When I consider the short duration of my life, swallowed up in the eternity before and after, the little space which I fill, and even can see, engulfed in the infinite immensity of spaces of which I am ignorant, and which know me not, I am frightened, and am astonished at being here rather than there; for there is no reason why here rather than there, why now rather than then. Who has put me here? By whose order and direction have this place and time been allotted to me? . . . .

#206 (Brunschvicg) = #201 (Lafuma)

The eternal silence of these infinite spaces frightens me.

[9]This and the following quotations are from Pascal's *Pensées* as translated by W. F. Trotter, which translation cites each item by the number in the ordering of Léon Brunschvicg. The numberings of the *pensées* assigned in the ordering of Louis Lafuma are also provided.

# 242 (Brunschvicg) = #781 (Lafuma)

I admire the boldness with which these persons undertake to speak of God. In addressing their argument to infidels, their first chapter is to prove Divinity from the works of nature. I should not be astonished at their enterprise, if they were addressing their argument to the faithful; for it is certain that those who have the living faith in their heart see at once that all existence is none other than the work of the God whom they adore. But for those in whom this light is extinguished, and in whom we purpose to rekindle it, persons destitute of faith and grace, who, seeking with all their light whatever they see in nature that can bring them to this knowledge, find only obscurity and darkness; to tell them that they have only to look at the smallest things which surround them, and they will see God openly, to give them, as a complete proof of this great and important matter, the course of the moon and planets, and to claim to have concluded the proof with such an argument, is to give them ground for believing that the proofs of our religion are very weak. And I see by reason and experience that nothing is more calculated to arouse their contempt.

#346 (Brunschvicg) = #759 (Lafuma)
Thought constitutes the greatness of man.

#347 (Brunschvicg) = #200 (Lafuma)

Man is but a reed, the most feeble thing in nature; but he is a thinking reed. The entire universe need not arm itself to crush him. A vapour, a drop of water suffices to kill him. But, if the universe were to crush him, man would still be more noble than that which killed him, because he knows that he dies and the advantage which the universe has over him; the universe knows nothing of this.

All our dignity consists, then, in thought. By it we must elevate ourselves, and not by space and time which we cannot fill. Let us endeavour then to think well; this is the principle of morality.

#348 (Brunschvicg) = #113 (Lafuma)

*A thinking reed.*—It is not from space that I must seek my dignity, but from the government of my thought. I shall have no more if I possess worlds. By space the universe encompasses and swallows me up like an atom; by thought I comprehend the world.

Thomas Traherne (1637–1674)

From "Insatiableness," II

'Tis mean Ambition to define
A single World:
To many I aspire,
Tho one upon another hurl'd:
Nor will they all, if they be all confin'd,
Delight my Mind.

This busy, vast, enquiring Soul
Brooks no Controul:
'Tis very curious too.
Each one of all those Worlds must be
Enricht with infinit Variety
And Worth; or 'twill not do.

John Milton (1608–1674)

From *Paradise Lost* (1667)

He scarce had ceas't when the superiour Fiend
Was moving toward the shore; his [Satan's] ponderous shield,
Ethereal temper, massy, large and round,
Behind him cast; the broad circumference
Hung on his shoulders like the Moon, whose Orb
Through Optic Glass the *Tuscan* Artist [Galileo] views
At Ev'ning from the top of *Fesole*,
Or in *Valdarno*, to descry new Lands,
Rivers or Mountains in her spotty Globe.

Book I, lines 283–91

. . .As when by night the Glass
Of *Galileo*, less assur'd, observes
Imagined Lands and Regions in the Moon. . . .

Book V, lines 261–3

[Adam to Raphael]
"When I behold this goodly Frame, this World,
Of Heav'n and Earth consisting, and compute
Thir magnitudes, this Earth a spot, a grain,
An Atom, with the Firmament compar'd
And all her number'd Stars, that seem to roll
Spaces incomprehensible (for such
Thir distance argues and thir swift return
Diurnal) merely to officiate light
Round this opacous Earth, this punctual spot,
One day and night; in all thir vast survey
Useless besides; reasoning I oft admire
How Nature wise and frugal could commit
Such disproportions, with superfluous hand
So many nobler Bodies to create,
Greater so manifold to this one use,
For aught appears, and on thir Orbs impose
Such restless revolution day by day
Repeated. . . ."
. . . . . . . . . . . . . . . . . . . . . . . . . . . . . . . . .
And *Raphael* now to *Adam's* doubt propos'd
Benevolent and facile thus repli'd:
"To ask or search I blame thee not; for Heav'n
Is as the Book of God before thee set,
Wherein to read his wond'rous Works , and learn
His Seasons, Hours, or Days, or Months, or Years;
This to attain, whether Heav'n move or Earth,
Imports not, if thou reck'n right; the rest
From Man or Angel the great Architect
Did wisely to conceal, and not divulge
His secrets, to be scann'd by them who ought
Rather admire; or if they list to try
Conjecture, he his Fabric of the Heav'ns
Hath left to thir disputes, perhaps to move
His laughter at thir quaint Opinions wide
Hereafter, when they come to model Heav'n,
And calculate the Stars; how they will wield
The mighty frame, how build, unbuild, contrive
To save appearances; how gird the Sphere

With Centric and Eccentric scribbl'd o'er,
Cycle and Epicycle, Orb in Orb. . . .
. . . . . . . . . . . . . . . . . . . . . . . . . . . . . . . .
God to remove his ways from human sense,
Plac'd Heav'n from Earth so far, that earthly sight,
If it presume, might err in things too high,
And no advantage gain. What if the Sun
Be Centre to the World, and other Stars
By his attractive virtue and their own
Incited, dance about him various rounds?
Thir wandring course now high, now low, then hid,
Progressive, retrograde, or standing still,
In six thou seest, and what if sev'nth to these
The Planet Earth, so steadfast though she seem,
Insensibly three different Motions move?
Which else to several Spheres thou must ascribe,
Mov'd contrary with thwart obliquities,
Or save the Sun his labour, and that swift
Nocturnal and Diurnal rhomb suppos'd
Invisible else above all Stars, the Wheel
Of Day and Night; which needs not thy belief,
If Earth industrious of herself fetch Day
Travelling East, and with her part averse
From the Suns beam meet Night, her other part
Still luminous by his ray. What if that light,
Sent from her through the wide transpicuous air,
To the terrestrial Moon be as a star,
Enlightning her by Day, as she by Night
This Earth? reciprocal, if Land be there,
Fields and Inhabitants: Her spots thou seest
As Clouds, and Clouds may rain, and Rain produce
Fruits in her soft'nd Soil, for some to eat
Allotted there; and other Suns perhaps,
With thir attendant Moons thou wilt descry
Communicating Male and Female Light,
Which two great Sexes animate the World,
Stor'd in each Orb perhaps with some that live.
For such vast room in Nature unpossest
By living Soul, desert and desolate,

Only to shine, yet scarce to contribute
Each Orb a glimpse of Light, convey'd so far
Down to this habitable, which returns
Light back to them, is obvious to dispute.

. . . . . . . . . . . . . . . . . . . . . . . . . . . . . . . . . . .

Dream not of other Worlds, what Creatures there
Live, in what state, condition or degree
Contented that thus far hath been reveal'd
Not of Earth only but of highest Heav'n."

Book VIII, lines 15–32; 64–84; 119–58; 175–8

Some say he bid his Angels turn askance
The Poles of Earth twice ten degrees and more
From the Suns Axle; they with labour push'd
Oblique the Centric Globe: Some say the sun
Was bid turn Reins from th' Equinoctial Road
Like distant breadth to *Taurus* with the Sev'n
*Atlantic* Sisters, and the *Spartan* Twins,
Up to the *Tropic* Crab; thence down amain
By *Leo* and the *Virgin* and the *Scales*,
As deep as *Capricorn,* to bring in change
Of Seasons to each Clime; else had the Spring
Perpetual smil'd on Earth with vernant Flow'rs,
Equal in Days and Nights, except to those
Beyond the Polar Circles. . . .

Book X, lines 668–81

*Eighteenth Century*

Gottfried Wilhelm Leibniz (1646–1716)

From his *Théodicée* (1710)
. . . one may imagine possible worlds without sin and without unhappiness . . . but these same worlds again would be very inferior to ours in goodness. I cannot show you this in detail. For can I know and can I present infinities to you and compare them together? But you must judge with me *ab effectu,* since God has chosen this world as it is. We know . . . that often an evil brings forth a good. . . .[10]

From his *Nouveaux essais sur l'entendement humain (New Essays Concerning Human Understanding)* (Published 1765)
If . . . someone came from the moon . . ., we would take him to be a lunarian; and yet we might grant him . . . the title *man* . . .; but if he asked to be baptized, and to be regarded as a convert to our faith, I believe that we would see great disputes arising among the theologians. And if relations were opened up between ourselves and these planetary men—whom M. Huygens says are not much different from men here—the problem would warrant calling an Ecumenical Council to determine whether we should undertake to propagate of the faith in regions beyond our globe. No doubt some would maintain that rational animals from those lands, not being descended from Adam, do not partake of redemption by Jesus Christ. . . . Perhaps there would be a majority decision in favour of the safest course, which would be to baptize these suspect humans conditionally. . . . But I doubt they would ever be found acceptable as priests of the Roman Church, because until there was some revelation their consecrations would always be suspect. . . . [T]hese bizarre fictions have their uses in abstract studies, as aids to a better grasp of the nature of our ideas.[11]

---

[10]G. W. Leibniz, *Theodicy: Essays upon the Goodness of God, the Freedom of Man, and the Origin of Evil*, trans. E. M. Huggard (New Haven, 1952), p. 129.

[11]G. W. Leibniz, *New Essays on Human Understanding*, trans. Peter Remnant and Jonathan Bennett (Cambridge, England, 1981), p. 314.

## Joseph Addison (1672–1719)

"Ode" from "Spectator #465" (1712), written with reference to Psalm 19

The Spacious Firmament on high,
With all the blue Etherial sky,
And spangled Heav'ns, a Shining Frame,
Their great Original proclaim:
Th'unwearied Sun, from day to day
Does his Creator's Pow'r display,
And publishes to every Land
The Work of an Almighty Hand.

Soon as the Evening Shades prevail,
The Moon takes up the wondrous Tale,
And nightly to the listning Earth
Repeats the Story of her Birth,
Whilst all the Stars that round her burn
And all the Planets, in their turn,
Confirm the Tidings as they rowl,
And spread the Truth from Pole to Pole.

What though, in solemn Silence, all
Move round the dark terrestrial Ball?
What tho' nor real Voice nor Sound
Amid their radiant Orbs be found?
In Reason's Ear they all rejoice
And utter forth a glorious Voice,
For ever singing, as they shine,
'The Hand that made us is Divine.'

## Alexander Pope (1688–1744)

From *Essay on Man* (1733–34)

Thro' worlds unnumber'd tho' the God be known,
'Tis ours to trace him only in our own.
He, who thro' vast immensity can pierce,
See worlds on worlds compose one universe,
Observe how system into system runs,

What other planets circle other suns,
What vary'd Being peoples ev'ry star,
May tell why Heav'n has made us as we are.

Epistle I, lines 21–8

Know then thyself, presume not God to scan;
The proper study of Mankind is Man.
Plac'd on this isthmus of a middle state,
A Being darkly wise, and rudely great:
With too much knowledge for the Sceptic side;
With too much weakness for the Stoic's pride,
He hangs between. . . .

Epistle II, lines 1–7

Edward Young (1683–1765)

From *Night Thoughts* (1742–45), "Night the Ninth"
This gorgeous apparatus! this display!
This ostentation of creative power!
This theatre!—what eye can take it in?
By what divine enchantment was it raised,
For minds of the first magnitude to launch
In endless speculation, and adore?
One sun by day, by night ten thousand shine;
And light us deep into the Deity;
How boundless in magnificence and might!
. . . . . . . . . . . . . . . . . . . . . . . . . . . . . . . .
Devotion! daughter of Astronomy!
An undevout astronomer is mad.
True; all things speak a God; but in the small,
Men trace out Him; in great, He seizes man:
Seizes, and elevates, and wraps, and fills
With new inquiries, 'mid associates new.
Tell me, ye stars! ye planets! tell me, all
Ye starred, and planeted, inhabitants! . . .
. . . . . . . . . . . . . . . . . . . . . . . . . . . . . . . .
. . .what swarms
Of worlds, that laugh at earth! immensely great!
Immensely distant from each other's spheres!

What, then, the wondrous space through which they roll?
At once it quite engulfs all human thought:
'Tis comprehension's absolute defeat. . . .
. . . . . . . . . . . . . . . . . . . . . . . . . . . . . .
[Addressed to extraterrestrials]
Whate'er your nature, this is past dispute,
Far other life you live, far other tongue
You talk, far other thought, perhaps, you think,
Than man. How various are the works of God?
But say, what thought? Is Reason here enthroned
And absolute? Or Sense in arms against her?
Have you two lights? or need you no reveal'd? . . .
And had your Eden an abstemious Eve? . . .
Or if your mother fell, are you redeem'd? . . .
Is this your final residence? If not,
Change you your scene, translated? or by death?
And if by death: what death? . . .

Samuel Pye

From his *Moses and Bolingbroke* (1765), pp. 60–63
GENESIS, Chap. I
1. In the beginning God created the heaven and Jupiter.
. . . . . . . . . . . . . . . . . . . . . . . . . . . . . .
16. And God made five great lights; the greater light to rule the day, and the lesser lights to rule the night: and stars also.

Chap. II
2. And on the fifteenth day God ended his work which he had made: and he rested on the fifteenth day from all his work, which he had made
3. And God blessed the fifteenth day, and sanctified it. . . .

Immanuel Kant (1724–1804)

From Kant's *Kritik der reinen Vernunft* (1781), Preface to the second edition (1787)
We here propose to do just what Copernicus did in attempting to explain the celestial movements. When he found that he could

make no progress by assuming that all the heavenly bodies revolved around the spectator, he reversed the process, and tried the experiment of assuming that the spectator revolved, while the stars remained at rest. We may make the same experiment with regard to the intuition of objects.[12]

Pierre Simon Laplace (1749–1827)

From Laplace's *Exposition du Système du Monde* (1796)

Astronomy considered in its entirety is the finest monument of the human mind, the noblest essay of its intelligence. Seduced by the illusions of the senses and of self-pride, for a long time man considered himself as the centre of the movement of the stars; his vainglory has been punished by the terrors which its own ideas have inspired. At last the efforts of several centuries brushed aside the veil which concealed the system of the world. We discover ourselves upon a planet, itself almost imperceptible in the vast extent of the solar system, which in its turn is only an insensible point in the immensity of space. The sublime results to which this discovery has led should suffice to console us for our extreme littleness, and the rank which it assigns to the earth. Let us treasure with solicitude, let us add to as we may, this store of higher knowledge, the most exquisite treasure of thinking beings.

[12]Immanuel Kant, *Critique of Pure Reason*, trans. J. M. D. Meiklejohn (London, 1934), p. 12.

*Nineteenth and Twentieth Century*

Johann Wolfgang von Goethe (1749–1832)

From his *Theory of Colours* (1810)
Humanity has perhaps never faced a greater challenge; for by [Copernicus's] admission [that humanity is not the center of the universe], how much else did not collapse in dust and smoke: a second paradise, a world of innocence, poetry and piety, the witness of the senses, the conviction of a religious and poetic faith . . . ; no wonder that men had no stomach for all this, that they ranged themselves in every way against such a doctrine. . . .[13]

Authorship Unknown

From "The Astronomer's Drinking Song" in Augustus De Morgan's *Budget of Paradoxes* (1866)
Whoe'er would search the starry sky,
  Its secrets to divine, sir,
Should take his glass—I mean, should try
  A glass or two of wine, sir!
True virtue lies in golden mean,
  And man must wet his clay, sir;
Join these two maxims, and 'tis seen
  He should drink his bottle a day, sir!
. . . . . . . . . . . . . . . . . . . . . . . . . . . . . . .
When Ptolemy, now long ago,
  Believed the earth stood still, sir,
He never would have blundered so,
  Had he but drunk his fill, sir:
He'd then have felt it circulate,
  And would have learnt to say, sir,
The true way to investigate
  Is to drink your bottle a day, sir!

Copernicus, that learned wight,
  The glory of his nation,

[13] J. W. Goethe, *Zur Farbenlehre* in Goethe, *Sämtliche Werke*, vol. 40 (Stuttgart, 1902), p. 185.

With draughts of wine refreshed his sight,
And saw the earth's rotation;
Each planet then its orb described,
The moon got under way, sir;
These truths from nature he imbibed
For he drank his bottle a day, sir!

The noble Tycho placed the stars,
Each in its due location;
He lost his nose by spite of Mars,
But that was no privation:
Had he but lost his mouth, I grant
He would have felt dismay, sir,
Bless you! he knew what he should want
To drink his bottle a day, sir!

Cold water makes no lucky hits;
On mysteries the head runs:
Small drink let Kepler time his wits
On the regular polyhedrons:
He took to wine, and it changed the chime,
His genius swept away, sir,
Through area varying as the time
At the rate of a bottle a day, sir!

Poor Galileo, forced to rat
Before the Inquisition,
E pur si muove was the pat
He gave them in addition:
He meant, whate'er you think you prove,
The earth must go its way, sirs;
Spite of your teeth I'll make it move,
For I'll drink my bottle a day, sirs!

Otto Neugebauer

From his *Exact Sciences in Antiquity* (1957)
The popular belief that Copernicus's heliocentric system constitutes a significant simplification of the Ptolemaic system is obviously wrong. The choice of the reference system has no

effect whatever on the structure of the model, and the Copernican models themselves require about twice as many circles as the Ptolemaic models and are far less elegant and adaptable.[14]

Michael Polanyi (1891–1976)

From *Personal Knowledge* (1958)

What is the true lesson of the Copernican revolution? Why did Copernicus exchange his actual terrestrial station for an imaginary solar standpoint? The only justification for this lay in the greater intellectual satisfaction he derived from the celestial panorama as seen from the sun instead of the earth. Copernicus gave preference to man's delight in abstract theory, at the price of rejecting the evidence of our senses, which present us with the irresistible fact of the sun, the moon, and the stars rising daily in the east to travel across the sky towards their setting in the west. In a literal sense, therefore, the new Copernican system was as anthropocentric as the Ptolemaic view, the difference being merely that it preferred to satisfy a different human affection.

It becomes legitimate to regard the Copernican system as more objective than the Ptolemaic only if we accept this very shift in the nature of intellectual satisfaction as the criterion of greater objectivity. This would imply that of the two forms of knowledge, we should consider as more objective that which relies to a greater measure on theory rather than on more immediate sensory experience. So that, the theory being placed like a screen between our senses and the things of which our senses otherwise would have gained a more immediate impression, we would rely increasingly on theoretical guidance for the interpretation of our experience, and would correspondingly reduce the status of our raw impressions to that of dubious and possibly misleading appearances.

. . . . . . . . . . . . . . . . . . . . . . . . . . . . . . . .

Here, then, are the true characteristics of objectivity as exemplified by the Copernican theory. Objectivity does not demand that we estimate man's significance in the universe by the minute size of his body, by the brevity of his past history or his

---

[14]Otto Neugebauer, *The Exact Sciences in Antiquity*, 2nd ed. (Providence, RI, 1957), p. 204.

probable future career. It does not require that we see ourselves as a mere grain of sand in a million Saharas. It inspires us, on the contrary, with the hope of overcoming the appalling disabilities of our bodily existence, even to the point of conceiving a rational idea of the universe which can authoritatively speak for itself. It is not a counsel of self-effacement, but the very reverse—a call to the Pygmalion in the mind of man.[15]

[15]Michael Polanyi, *Personal Knowledge: Towards a Post-Critical Philosophy* (New York, 1964), pp. 3–5.

# Appendix

## *Archaeoastronomy*

**Stonehenge as Seen from the Heel Stone**

### *Introduction*

This appendix consists of a brief introduction to **archaeoastronomy**. Although its focus is **Stonehenge**, the most famous of the nearly one thousand **megalithic** (large stone) sites scattered over the British Isles, the principles involved in investigating whether Stonehenge has an astronomical orientation are applicable to studies of these other sites, hundreds of which have been analyzed. In studying Stonehenge, we shall be looking at developments from about four thousand years ago. Ancient as the megaliths are, the modern claims that they have astronomical functions date primarily from the 1960s. We shall consequently be dealing simultaneously with very recent and very ancient materials.

Stonehenge, which ranks second only to the Tower of London among Britain's most popular tourist sites, is visited by over a million persons each year. Visitors easily see the monument as an impressive feat of elementary engineering, or possibly of slave labor, yet few possess sufficient knowledge of astronomy to assess claims as to its astronomical orientation. If seen in the latter way, Stonehenge appears as a creation of preliterate although not prescientific genius. The goal of these materials is to provide a basis for assessing these claims. If they are correct, Stonehenge contains a message, written not in letters but in the positions of stones, not in words but in mathematics.

*Description of Stonehenge and History of Its Construction*

Stonehenge was not built all at once; in fact, over a thousand years passed from its first establishment to its completion. In a sense, there are three Stonehenges, or were at least three major phases in its construction. Consequently, archaeoastronomers refer to Stonehenge I, II, and III and portray each of them as quite different from the others.

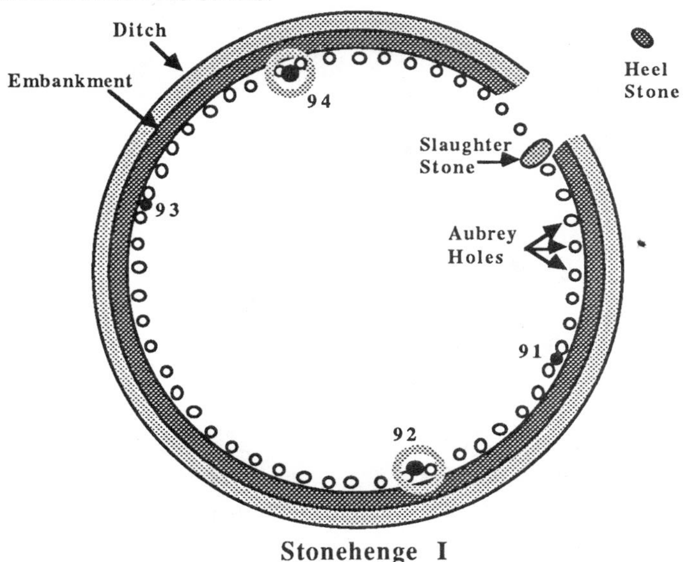


**Stonehenge I**

The construction of Stonehenge I began around 2800 B.C. with a **circular embankment** 6 feet in height and about 320 feet in diameter. The materials used in its construction came from the ditch that surrounds it. Toward the northeast there is a 35 foot gap in the embankment, which opening served as a sort of entrance. If one proceeds from the center out toward this gap, one comes first to the **Slaughter Stone,** which was probably once vertical. It has nothing to do with slaughtering. Farther out stands the **Heel Stone** (16 feet high). Within the circle at its edges are four **Station Stones** (designated by the numbers 91, 92, 93, 94), two of which (92 and 94) are on raised mounds. Running around the inside of the bank are 56 holes known as the **Aubrey Holes**. These were filled up in ancient times and their function is much disputed.

During the Stonehenge II phase, which dates from around 2100 B.C., two circles of relatively small blue stones were put into position.

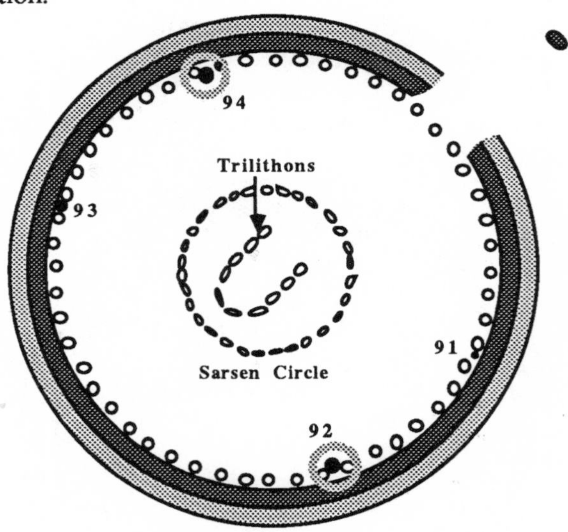


**Stonehenge III**

At the Stonehenge III stage, which is assigned to the period beginning around 2000 B.C., the **Sarsen Circle** of 30 upright sandstone blocks (of the type called Sarsen) was erected. These 30 standing stones, each weighing about 26 tons and standing 13 feet high, form a circle of 97.5 feet diameter. Inside the Sarsen Circle are five **trilithons**, arranged in a horseshoe shape, opening toward the Heel Stone. These stones are 20 to 25 feet high and weigh about 50 tons. Horizontal **lintels** lie atop the stones in both groups, held in place by a mortise and tenon arrangement.

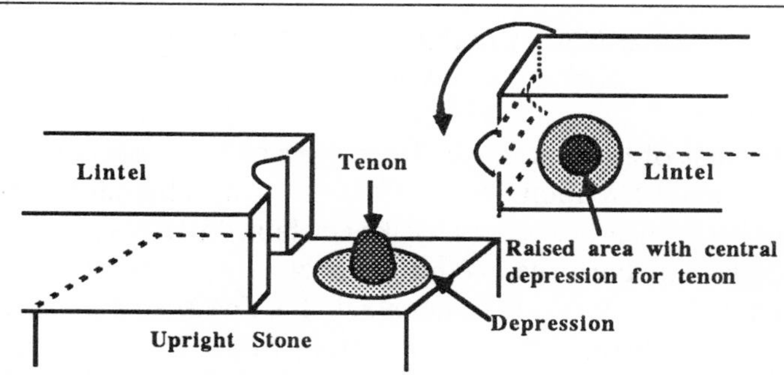


Many of the uprights and lintels have now fallen. A so-called **Altar Stone** is at the center of the horseshoe. At a second stage in the development of Stonehenge III, the x and y holes were positioned between the ring of Aubrey Holes and the Sarsen Circle. At the final stage, which dates from around 1550 B.C., rings of small bluestones were erected within the horseshoe and between it and the Sarsen Circle. There are other features of Stonehenge, but these are the ones of greatest interest.

The final overall structure of Stonehenge is represented in the next diagram.

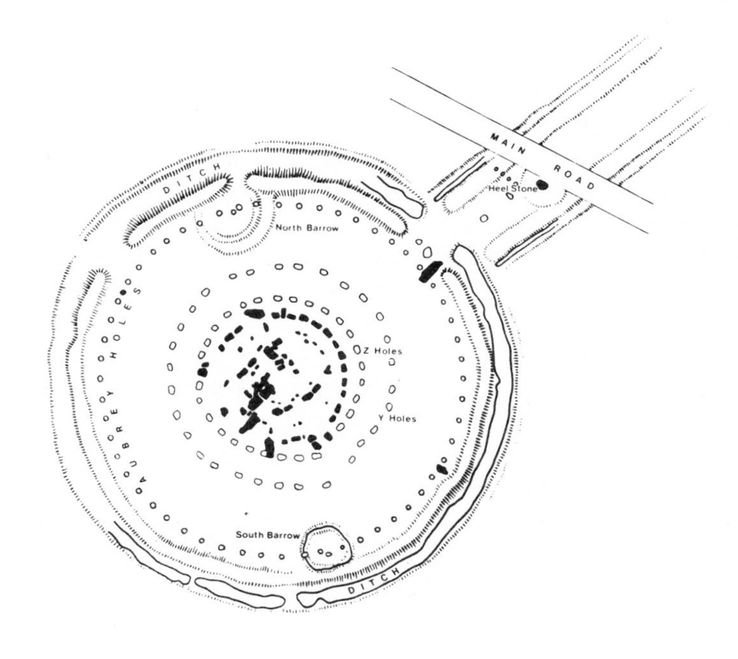

### *Brief History of the Study of Stonehenge, Especially Its Astronomical Functions*

Because of its impressive size and character, Stonehenge has long attracted attention. Such artists as Constable and Turner included it in their landscapes; Blake celebrated it in his poetry and drawings. In the seventeenth century, a number of authors studied it. Inigo Jones, for example, in his *The Most Remarkable Antiquity of Britain, Vulgarly Called Stone-Heng, Restored* (1655) argued against the legend presented by Geoffrey of Monmouth in his twelfth-century *History of the Kings of Britain* that Merlin had arranged to have the stones brought from Ireland and erected in England. Jones urged that they were erected by the Romans. Walter Charleton in his 1663 *Chorea Gigantum* argued for a Danish origin. The prominent antiquarian John Aubrey (1626–1697) in his *Monumenta Britannia* (which was not published at that time) described Stonehenge and reported his discovery of what are now called the "Aubrey holes." He attributed the structure to the Druids, who were supposed to have been an ancient order of Celtic priests. Although this claim was essentially baseless, the Druidic theory of Stonehenge had a long life.

In the eighteenth century, William Stukeley (1687–1765) perpetuated the Druid idea in his *Stonehenge, A Temple Restored to the British Druids* (1740). Stukeley's writings, however, contain the important claim that the main axis of the monument points to summer solstice sunrise. In 1771, Dr. John Smith published a work entitled *Choir Gaur the Grand Orrery of the Ancient Druids Called Stonehenge, Astronomically Explained and Proved to Be a Temple for Observing the Motions of the Heavenly Bodies*. Herein he asserted: ". . . I am convinced it to be an astronomical temple." He developed this idea in some detail, attributing a calendric function to Stonehenge.

In the nineteenth century, Rev. Edward Duke in an 1846 study, *The Druidic Temples of the County of Wilts*, noted that an alignment from mound 92 to station stone 91 points to summer solstice sunrise. Similarly, if one looked from mound 94 toward station stone 93, one would see the point on the horizon of winter solstice sunset. In 1880, the archaeologist Flinders Petrie published his *Stonehenge: Plans, Descriptions and Theories.* This included the most accurate and detailed survey of the monument made up to that time. No further developments occurred until early in the twentieth century, when Sir J. Norman Lockyer became involved in the study of temples and megaliths.

### *Lockyer*

Sir J. Norman Lockyer (1836–1920) was not only a prominent astrophysicist and the editor for half a century of the scientific journal *Nature,* but also an important pioneer of archaeoastronomy. While traveling in Greece in the early 1890s, Lockyer became interested in whether Greek and Egyptian monuments have astronomical orientations. Collaborating with him over the next two decades was a British architect turned archaeologist, F. C. Penrose. Lockyer energetically organized precise surveys of various Egyptian pyramids and other structures, publishing his results in his 1894 *Dawn of Astronomy.* Herein he set forth evidence that temples at Karnak were oriented toward the solstices, whereas at Giza they pointed toward the equinoctial sun. One of his more dramatic efforts was an attempt to date the construction of various Egyptian monuments by using data on the **changing obliquity of the ecliptic**. By calculating when the solstitial sun was at the point to which the shrine of Amen-Ra at Karnak was aligned, he estimated the date of its erection as 3700 B.C. Lockyer also claimed that many Egyptian monuments were oriented toward the heliacal rising of such stars as Sirius, gamma Draconis, and Canopus. The **heliacal rising** of a star occurs when it rises just before sunrise. He found that the temple of Isis at Denderah was oriented toward Sirius, the heliacal rising of which could be employed to predict the flooding of the Nile. He used the precessional movement of the stars to date the erection of temples oriented to stars.

In 1901, Lockyer, in collaboration with Penrose, began to study Stonehenge and eventually some other British megalithic sites. In 1906, he published his *Stonehenge and Other British Monuments Astronomically Considered.* Lockyer attempted to use astronomical information to date Stonehenge, arriving at the date 1680 B.C., give or take about 200 years. The method by which he did this depended upon the changing obliquity of the ecliptic, i.e., on the very slowly changing angle between the celestial equator and the ecliptic. This change occurs at the rate of about 1 degree per 4000 years. By comparing the orientation of Stonehenge with values of this inclination calculated for different periods, Lockyer could estimate the time of its erection. In attempting this, he faced a number of problems; for example, what should be taken as defining the orientation of Stonehenge? Should it be:

1. From the center of the stone circle to the Heel Stone?
2. From the center of the monument to the center of the avenue?

or

3. From the center of the monument to some distant location?

Lockyer chose the last method, using a marker 8 miles distant. Another question was: how should one specify the point of sunrise? Again there are at least 3 possibilities:

1. The point at which the sun is tangent to the horizon,
2. The point where the sun is bisected by the horizon, and
3. The point of the first or last gleam as the sun rises or sets.

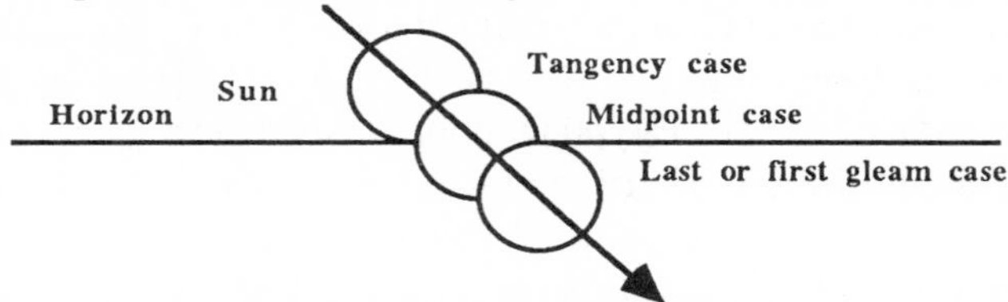


Lockyer seems to have chosen the third definition. All these, and other factors, show that this dating technique is substantially less precise than it might at first seem.

Lockyer missed a number of significant Stonehenge alignments later found; he did, however, claim alignments for some of the quarter days—between solstice and equinox. He claimed some stellar alignments, but did not carefully investigate lunar alignments. Moreover, he failed to use statistical techniques, which later came to be seen as necessary in analyzing alignment claims. After his book received a mixed reaction, probably more hostile than favorable, interest in archaeoastronomy declined until it was revived in the early 1960s by Gerald Hawkins.

*More on the Motions of the Moon*

As a preliminary to the presentation of the archaeoastronomical ideas of Hawkins and his successors, it is necessary to supplement the discussion of the motions of the moon provided in Chapter One, where lunar and solar eclipses as well as the draconitic period of the moon were discussed. It is first of all necessary to elaborate on the moon's draconitic period.

Understanding the moon's draconitic period can be facilitated by considering the next diagram, which represents the 5° inclination of the moon's orbit to the plane of the earth's motion.

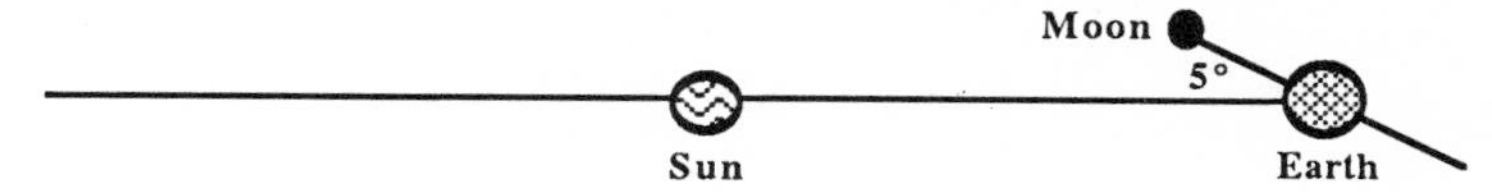


The moon's motion is actually more complex than this; in fact, the plane of the moon's orbit undergoes a slow rotation, moving in a direction opposite to that in which the moon moves. The period of this rotation is 18.61 years. This accounts for the difference between the moon's sidereal and draconitic periods. One result of this motion is that after half of the 18.61 year period, the relative orientation of the planes of the orbits will be that shown in the next diagram.

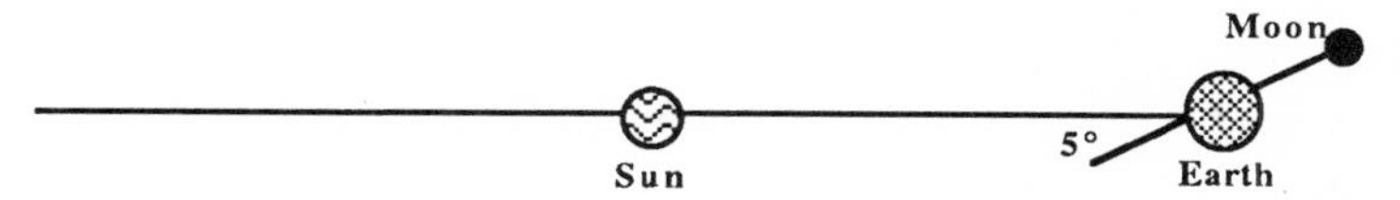


Another way of describing this phenomenon is to state that each of the moon's nodes makes a complete revolution through the moon's orbit in a period of 18.61 years, as shown in the next diagram.

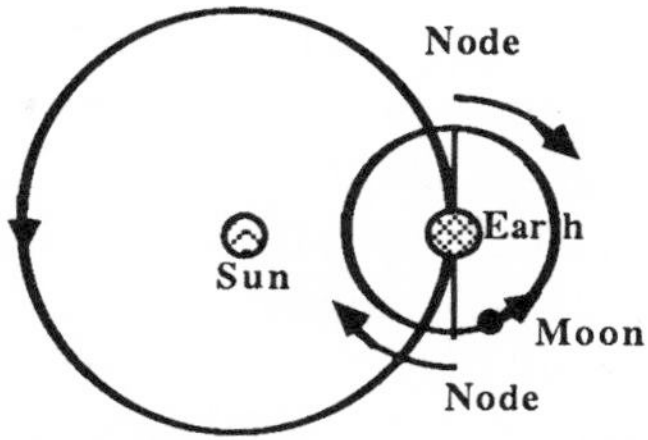


The 18.61 Year Circuit of the Moon's Nodes

This shift in the moon's nodes accounts for the difference between the moon's synodic and draconitic months.

The next diagram represents the moon's apparent path on the starry vault at a time when the moon's ascending node coincides with the vernal equinox and the moon's descending node coincides with the autumnal equinox. For this configuration, the moon will be seen to move as far as 28.5° (23.5° + 5°) from the celestial equator. Because eclipses occur only when the moon is near or at one of the nodes of its orbit, it is important in predicting eclipses to know the position of the moon's nodes.

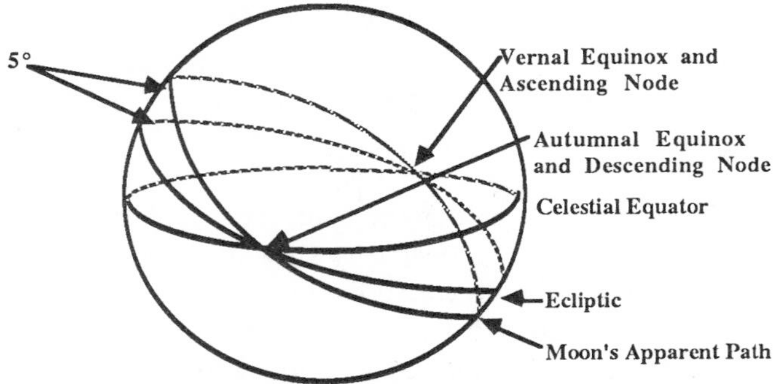


**Moon's Orbit at Greatest Inclination to the Celestial Equator with the Moon's Nodes at Vernal and Autumnal Equinox**

These concepts help explain an important phenomenon that is known as a **lunar standstill.** This term refers to the limiting values of angular distance of the moon from the celestial equator. From the previous diagram, it is evident that when the moon's nodes coincide with the equinoctial points, the moon can reach as far as 28.5° (23.5° + 5°) from the celestial equator, or it can come as near as 18.5° (23.5° – 5°) to the celestial equator. When the moon's nodes are not at the equinoctial points, its angular distance from the celestial equator will always be less than these extreme values. As the next diagram suggests, when the moon reaches these extreme values from the celestial equator, it tends to do so over an extended period. In other words, when during one lunar month the moon has reached this extreme value, it will continue to come very close to it over a number of succeeding months. This is shown by the flatness of the curve corresponding to the extreme distances (28.5° and 18.5°). The continuation of this phenomenon over a number of months is the source of the term standstill.

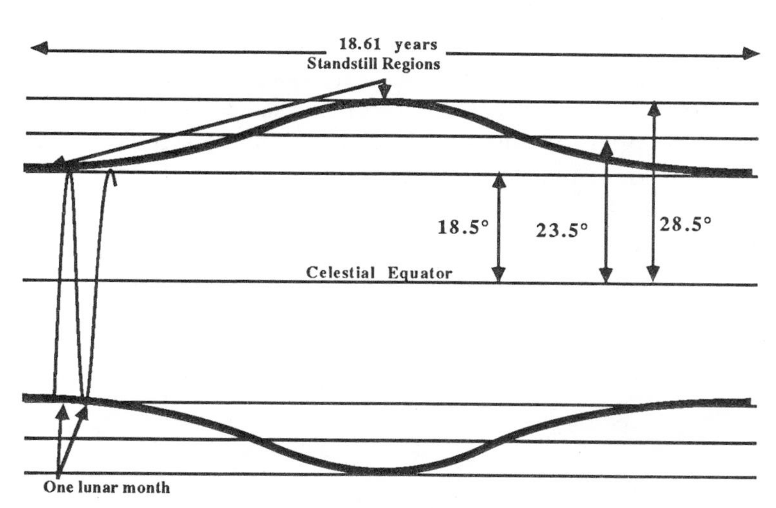


Variation in the Moon's Declination

When the moon during a standstill period continues to reach the maximum distance (28.5°) from the celestial equator, it is said to be at a **major lunar standstill;** when it persists at its least distance (18.5°), it is said to be at a **minor lunar standstill.**

Put differently and somewhat more simply, lunar observers at any location will be able to note that the moon at rising and setting will at various times reach an extremely northerly or southerly position and that it will persist in reaching these extreme positions over a matter of months. Moreover, it can be observed that the period between these occurrences is 18.61 years. The relevance of all this is that claims have been made that the megalith builders were intent on observing these extremum points and that a knowledge of the period separating these standstills would help them in predicting eclipses.

*Eclipse Prediction and Key Reference Points*

The occurrence of eclipses depends on a number of cycles, the most important of which are the moon's sidereal, synodic, and draconitic periods. As noted in Chapter One, prediction of solar eclipses is much more difficult than the prediction of lunar eclipses, the main reason being that lunar eclipses are seen over all parts of the earth where the moon is visible, whereas solar eclipses are visible only in the quite limited region of the earth that falls within the moon's shadow cone. That cone is relatively narrow because the sun and moon have nearly the same angular diameter.

A number of important results obtained thus far are represented in the next diagram, which shows fourteen key reference points on the horizon as they would be seen by a person on the equator.

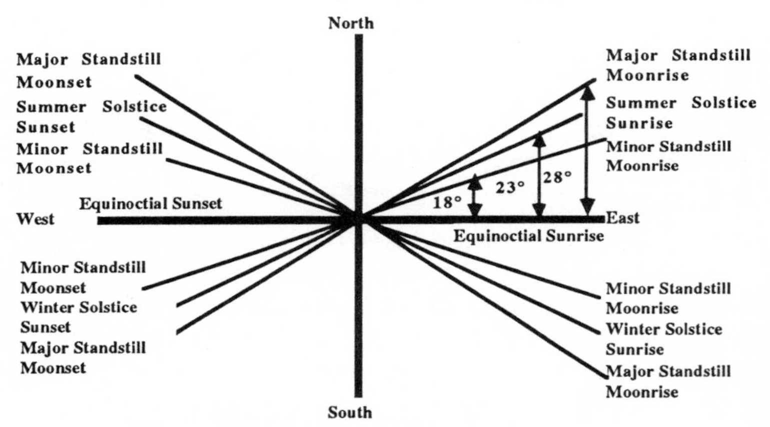


**Fourteen Key Celestial Points as Seen from the Equator**

This diagram shows where on the horizon a person located at the equator will see such events as Winter Solstice Sunrise. In that case it will be 23° to the south of due East. To select another case, the Summer Solstice Major Lunar Standstill will occur 28° to the north of due West. For persons located in more northerly regions, these angles will increase substantially. For example, at 51° north latitude, which is the latitude of Stonehenge, the solstices will appear about 40° on either side of the East-West line, whereas the standstills will appear about 10° farther to the south or north. Let us now turn to the claims that have been made that the megalith builders aligned their structures with a number of these key reference points, doing this partly to aid in eclipse prediction.

*Hawkins*

The contemporary burst of interest in archaeoastronomy is usually dated from October 26, 1963, when the British-born astronomer Gerald Hawkins, at that time teaching at Boston University, published a paper entitled "Stonehenge Decoded" in *Nature*, the journal that Lockyer had founded. This paper summarized the results of researches Hawkins had begun in 1960. What Hawkins did was to use less than one minute of computer time on the Smithsonian-Harvard computer to check for alignments between various features of Stonehenge and such astronomical horizon events as the risings and settings of prominent stars, the planets, the sun, and the moon. As he stated in his paper: "Stars and planets yielded no detectable correlation." For the sun and moon, however, he claimed a number of dramatic correlations, which are represented in the next diagram, which is based on the diagram in his "Stonehenge Decoded."

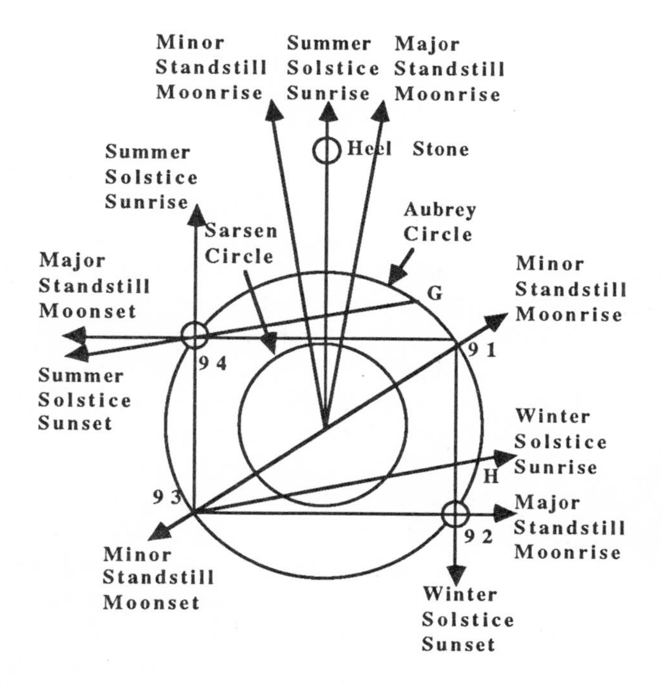


**Stonehenge Alignments Claimed by Hawkins**

Hawkins's results were derived on the basis of a number of assumptions, for example, about what alignments to check, whether the midpoint, lower edge, or last gleam of the sun should be used, and whether atmospheric refraction was the same amount then as now. In particular, Hawkins found the following solar correlations:

Summer Solstice Sunrise (Center to Heel Stone and 93 to 94)
Summer Solstice Sunset (G to 94)
Winter Solstice Sunrise (93 to H)
Winter Solstice Sunset (91 to 92)

For the moon, he investigated eight alignments, corresponding to the above positions $\pm$ 5°. In doing this he found six alignments. This was an exciting result, but it is important to note that in none of these alignments did the central Sarsen Circle or five trilithons play a part. Moreover, all these were alignments for what has come to be called Stonehenge I. He did find some alignments involving the Sarsen circle, but his results in this regard were overall rather less impressive. He detected no correlations with the Aubrey Holes.

### *Hawkins's Second Paper*

In the June 27, 1964 issue of *Nature,* Hawkins published a second, even more dramatic paper entitled "Stonehenge: A Neolithic Computer." In it, he asserted that "Stonehenge can be used as a digital computing machine," in particular, that it can be used to predict eclipses. Put differently, he claimed that he had found an astronomical function for the Aubrey holes. Recall the 56 Aubrey holes positioned at the edge of the circle. Hawkins noted that the period of the cycle for the return of the moon to the major standstill position is 18.61 years. This multiplied by 3 equals 18.61 x 3 = 55.83. In other words, there is a cycle of very nearly 56 years in the return of the moon to its standstill positions. Hawkins used that fact to suggest that if six stones were positioned at certain Aubrey holes and moved counterclockwise one hole per year, then the Stonehengers could have been predicted a large number of eclipses. Because this procedure is rather complex and because a simpler and less implausible procedure was soon proposed by the astronomer Fred

Hoyle (to be discussed shortly), Hawkins's method need not be discussed further.

In 1965 Hawkins published his conclusions in his book *Stonehenge Decoded*, which was written for the educated public. Around this time CBS devoted a special show to Hawkins, who described his results and even photographed the summer solstice sun rising over the Heel Stone. Hawkins had become a celebrity.

Problems, however, soon arose. The person who most vigorously and effectively called attention to these was the British archaeologist R. J. C. Atkinson, who was the leading expert on Stonehenge. In *Nature* for June 25, 1966, Atkinson reviewed Hawkins's book, describing it as "tendentious, arrogant, slipshod, and unconvincing." He urged, for example, referring to the Stonehenge I period, that it seemed improbable that "a barbarous and illiterate community" could achieve such wonders as Hawkins ascribed to them. Later that year he wrote a more lengthy review for *Antiquity*, using the title "Moonshine on Stonehenge." Although finding some merit in Hawkins's work, he pointed out numerous errors and problems. For example, he stated that the evidence was against Holes F, G, and H being man-made and stressed that the maps used by Hawkins were not detailed enough for his purposes. He also pointed out a number of mathematical and archaeological mistakes.

### *Hoyle*

In the July 30, 1966 issue of *Nature*, Fred Hoyle, a prominent British astronomer, came to the rescue, if not of Hawkins, at least of some of his ideas. Hoyle put forth an ingenious theory of the Aubrey holes, which has advantages over Hawkins's theory. Like Hawkins, Hoyle proposed a way in which these holes could be used for eclipse prediction. In order to understand Hoyle's theory of the use of the Aubrey holes in eclipse prediction, recall that for an eclipse to occur, the moon must be near a nodal point and also its nodal point must line up with the position of the sun. In the accompanying diagram, let the circle of the 56 Aubrey holes represent the ecliptic. If we place a marker in Aubrey hole #10 and move it counterclockwise at the rate of 2 holes every 13 days, it will move through the 56 holes in (13 x 56/2 =) 364 days or almost exactly one year. This marker will consequently effectively represent the position of the sun. Place another marker in alignment with the moon's position relative to the ecliptic and move it counterclockwise at the rate of two holes per day. It will complete the circuit in (56/2=) 28 days or about the sidereal period of the moon. Place a pair of markers opposite each other, say at holes #18 and #46. Move each marker clockwise at the rate of 1 hole every four months. Each will complete the circuit of 56 holes in (56/3 =) 18.67 years. If once aligned with the lunar nodes, these markers will come very close to preserving that alignment.

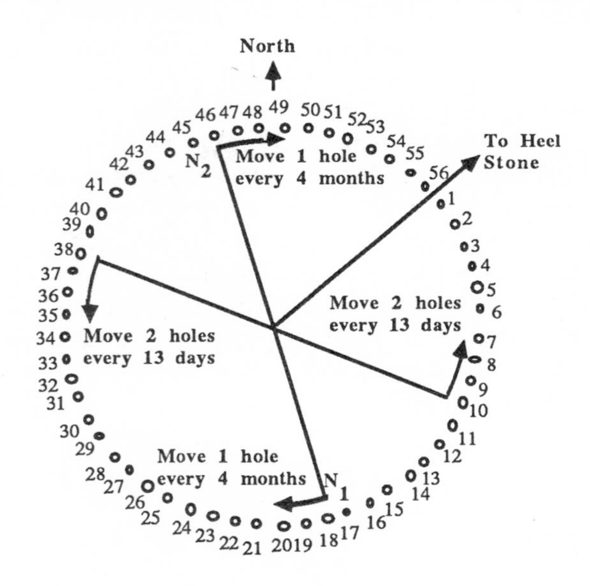


**Diagram Illustrating Hoyle's Suggestion on the Use of the Aubrey Holes for Eclipse Prediction**

Consequently, whenever the sun and moon markers are simultaneously near the nodal markers, an eclipse is very possible. The consensus is that Hoyle's method is probably beyond the sophistication of the Stonehengers. Hoyle presented his ideas more fully in his book *On Stonehenge*.

*Newham*

At the end of his review in *Antiquity* of Hawkins's book, Atkinson mentioned two other archaeoastronomers, Newham and Thom. The ideas of these fascinating figures, both of whom had been searching for astronomical alignments for some years before Hawkins's first paper, repay study. C. A. (Peter) Newham (d. 1974) was an engineer living in northern England when around 1959 he became interested in Stonehenge. An active amateur astronomer, Newham, after retiring in 1958, investigated Stonehenge for astronomical alignments. By 1962, he had achieved a number of results and received some encouragement from Atkinson. Nonetheless, *Antiquity* rejected a paper he had submitted. His ideas did attract the attention of the science writer for the *Yorkshire Post*, which on March 16, 1963, published an article on Newham's work. It is noteworthy that this was six months before the first paper by Hawkins. That article summarized Newham's conclusions in a diagram similar to the next diagram.

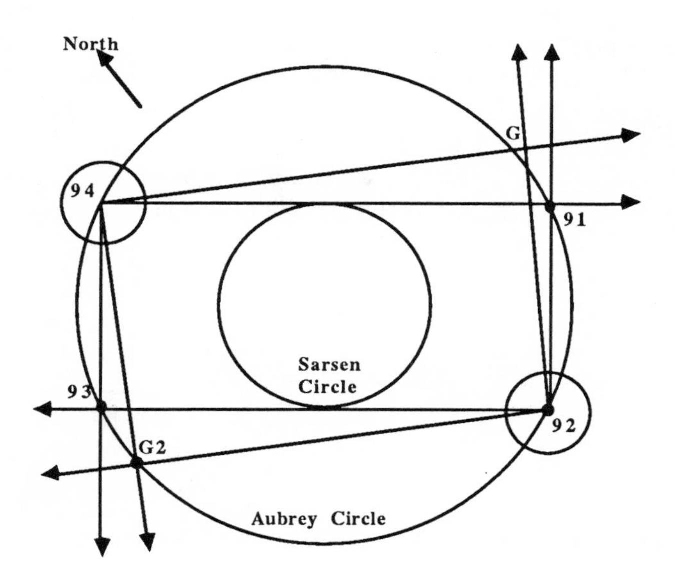


**Newham's Alignments as Given in the *Yorkshire Post***

As this diagram shows, Newham had claimed six alignments.

Summer Solstice Sunrise (92 to 91)
Winter Solstice Sunrise (94 to G)
Winter Solstice Sunset (94 to 93)
Northern Major Standstill Moonrise (92 to G)
Northern Major Standstill Moonset (92 to 93)
Southern Major Standstill Moonrise (94 to 91)

Moreover, Newham postulated a hole G2 that he believed would line up with 92 to give (92 to G2) Summer Solstice Sunset. Also 94 to G2 would give the Southern Major Standstill Moonset. Had evidence of G2 been found at Newham's postulated location, it would have given him all 4 solar solstice points and all 4 extreme moon positions, but no hole at G2 was found. We shall return to this problem shortly, but first it should be noted that Newham did arrange for the publication of his ideas in a pamphlet, which, because of a fire in the print shop, was published a few months after Hawkins's paper. Little attention, however, was paid to Newham's work, even after Hawkins's claims had made that astronomer famous.

Another interesting feature of the first report of Newham's work is that Newham provided an explanation of why the Sarsen Circle was centered on a point about a yard distant from the center of the Aubrey circle. Newham noted that were this not the case, it would have blocked the 92 to 93 alignment. It is significant too that Newham found that the best correlations were made when the first or last gleam of the sun or moon was used for sighting, whereas Hawkins had used the full orb.

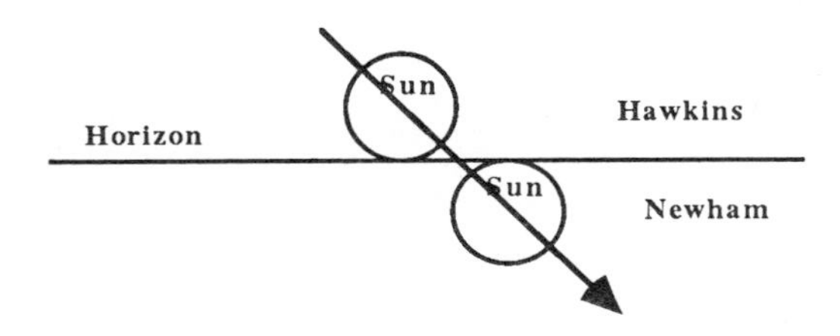


In 1966, Newham published a paper in *Nature* in which he provided an explanation of the about 50 **postholes** found in the causeway or opening in the mound surrounding Stonehenge. These, he suggested, were made to hold posts placed to mark the most northerly positions of the moonsets over various cycles.

As mentioned previously, Newham postulated a G2 hole, which he believed would give the Summer Solstice Sunset when viewed from 92 and the Southern Major Standstill Moonset when viewed from 94. G2 was never found, but in 1966, while the parking area at Stonehenge was being expanded, three large postholes were located to the northwest of Stonehenge at a distance of about 830 feet. The holes were about 2 1/2 feet in diameter. In his 1970 booklet, *The Astronomical Significance of Stonehenge*, Newham showed that were these holes to have held posts 30 feet tall, those posts would have made excellent sights (see diagram) for precise observation. In fact, he found for them the alignments:

Summer Solstice Sunset (91 to posthole 1)
Minor Lunar Standstill Moonset (Heel Stone to posthole 2)
Northern Major Standstill Moonset (92 to center posthole),
as well as two others of less significance.

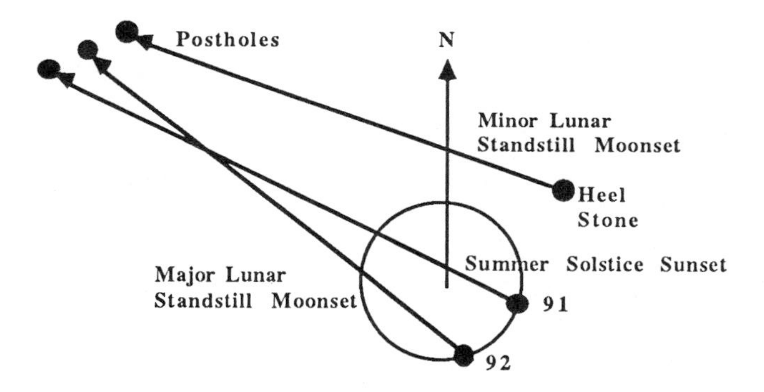


These alignments were very precise because of the distances of the holes from the observation points. In this booklet, he also made suggestions on the use of the Aubrey holes, blue stones, and the 30 Sarsen stones. These are viewed as more conjectural. Important as Newham was, his contributions to archaeoastronomy were far surpassed by those of another retiree, Alexander Thom, who emerged as the leading British scholar in this area.

*Thom*

Alexander Thom (1894–1985), who taught civil engineering at Oxford University from 1945 to his retirement in 1961, began surveying various megalithic sites in Britain at least as early as 1954, in which year he published a paper on megalithic solar observatories in Britain. He had become interested in megaliths in 1934 while sailing in Scotland. At that time he happened on the site at Callanish. Over the next four decades, he surveyed hundreds of sites and wrote three quite technical books as well as dozens of articles on megalithic astronomy. Some of his research and publication was done in collaboration with his son, Archibald S. Thom. Although Thom had been developing his theories long before Hawkins's book appeared, he had avoided Stonehenge, which he finally analyzed in the early 1970s. Before discussing his conclusions about Stonehenge, let us look at some of his other studies.

First of all, it is noteworthy that Thom stressed a comparative and a statistical approach. He scrutinized hundreds of sites and analyzed alignments at them in a statistical way, sometimes representing his results in elaborate graphs. The chief reason why statistical analyses are so important is that at any given megalithic site consisting of a dozen or more stones, many alignments can be detected. For example, at a site consisting of 10 stones arranged in a rough circle, each of the stones determines an alignment with 9 other stones, making a total of 90 possible alignments. The number of such alignments necessitates that at least some of them will be astronomically significant. The issue then becomes whether such significant alignments occur at a frequency greater than the law of averages would indicate. If one finds that a given astronomical alignment recurs at numerous sites, this constitutes important evidence that the site was designed in terms of that alignment. If, however, a particular astronomical alignment occurs at a low statistical frequency among sites, this disconfirms the claim that sites were established in terms of that alignment.

One example of the conclusions that Thom reached by using a statistical methodology is his idea, based on his measurements of the diameters of numerous stone circles, that the builders of the megaliths showed a preference for a certain distance that he called the "megalithic yard," which Thom specified as 2.72 feet. Moreover, he claimed that the diameters of many rings are integral multiples of what he called the "megalithic rod," which is equal to 2 1/2 megalithic yards. He also urged that some of the megalithic rings that are not circles or ellipses are figures compounded of circles and elliptical sections put together in such a way as to make their perimeters come out to an integral number of megalithic rods, whereas their radii are in integral megalithic yards. We need not follow this line of development further.

Some of Thom's most dramatic claims are found in a cluster of papers he published around 1971 dealing with the **Carnac** site in Brittany. In that region of France, the largest single megalith, **"Le grand menhir brisé,"** is located. It consists of a stone 70 feet long and weighing over 300 tons. It is broken into five pieces, one of which is missing. Thom theorized that this giant megalith was designed to serve as a foresight for various locations some miles away. In particular, he specified four or five possible locales from which Le grand menhir brisé could serve as a foresight for determining positions of lunar standstills.

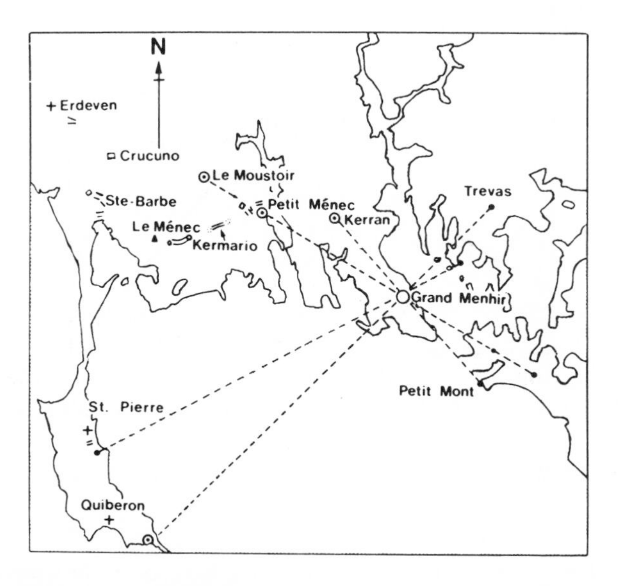


**Diagram Representing Thom's Claims for the Use of Le Grand Menhir Brisé as a Universal Foresight**

*Thom and Stonehenge*

It was with high expectations that persons interested in archaeoastronomy awaited Thom's study of Stonehenge. His papers appeared in 1974 and 1975 in the *Journal for the History of Astronomy*, which had become a leading repository for technical archaeoastronomy papers. In fact, one of its four issues each year is now devoted to that subject. Thom found at Stonehenge support for his theories of megalith mensuration; for example, the Sarsen stones are a megalithic rod wide and they are spaced 1/2 megalithic rod apart. Also he found confirmation for the Hawkins-Newham alignments, but what was most striking in his paper was his claim that four or possibly five locations at distances up to 9 miles away served as distant foresights (see diagram).

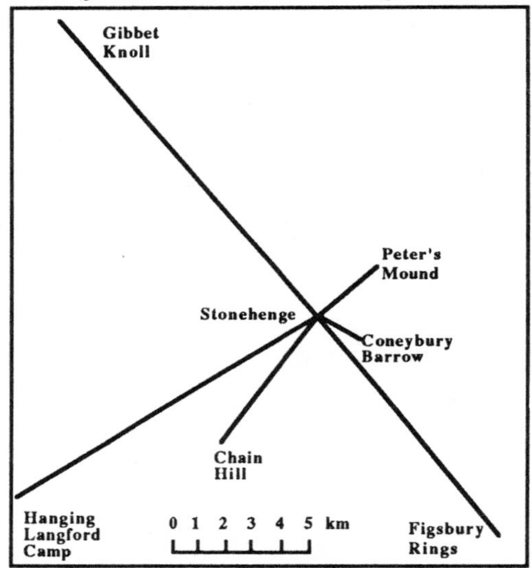


**Diagram of Thom's Hypothesis Concerning the Possibility of Distant Foresights at Stonehenge**

In concluding his second paper on Stonehenge, Thom asserted:

> We have shown that Stonehenge may have been the central point of a lunar observatory of a size comparable to the Observatory which we believe was centered on Le Grand Menhir Brisé in Brittany. In the Breton observatory there was

> a central foresight to be observed from the surrounding backsights, whereas at Stonehenge the observations were made from the monument to distant foresights.[1]

In this claim, Thom's word "may" deserves special emphasis. The evidence for the distant locations that he posited as foresights is decidedly slim. Moreover, some of his claims for an extremely high level of precision in megalithic observation have been effectively challenged. Nonetheless, his work is recognized as having brought a new level of precision to the study of the megaliths.

### *Conclusion*

The 1973 issue of *Current Anthropology* carried an interesting article by Elizabeth Baity entitled "Archaeoastronomy and Ethnoastronomy So Far," which included a bibliography of the subject consisting of over 500 items. Since that time, archaeoastronomical publications have appeared at the rate of fifty or more per year. This total includes articles not only on the megalithic sites, but also on Native American, Mayan, etc., astronomy as well as related areas of study. Some of this literature is excessively speculative and/or cultish. Much of it, however, is very carefully done and probably of lasting significance. If correct, the claims of such archaeoastronomers as Thom show that, even in pre-literate times, humans manifested strong interest in astronomy and developed sophisticated observational and computational techniques for dealing with the celestial motions. It is noteworthy that although much of archaeology involves reconstructing the most elemental practices of primitive groups—how their members dug, drank, killed, buried their dead, etc., archaeoastronomy provides an exciting new perspective by showing the intellectual heights humanity may have attained even at a very early period.

One of the most important directions that recent work in archaeoastronomy has taken is ever more detailed statistical

---

[1] Alexander Thom, Archibald Stevenson Thom, and Alexander Strang Thom, "Stonehenge as a Possible Lunar Observatory," *Journal for the History of Astronomy, 6* (1975), 29.

analyses. An example of such work is a book published in 1984 by Clive Ruggles in which he reported on a statistical study of 300 megalithic sites in western Scotland. In reviewing this book in 1985, Alvar Ellegård summarized Ruggles's conclusions in the following way:

> The alignments surveyed here are *not* random. There are clear clusterings, and the clusters can be connected with the movement of the rising and setting points of the moon along the horizon, and especially its extreme positions. There is somewhat weaker evidence for the winter solstice also being observed, but none, perhaps surprisingly, for the summer solstice. As for the equinoxes, the evidence is rather that the alignments avoided them.[2]

A comparable assessment of the situation was provided in 1988 by an astronomer, Ray Norris, in a survey of recent statistically oriented analyses in archaeoastronomy. After summarizing his conclusions and remarking that the statistical analyses he had discussed might lead one to think that "the astronomical hypotheses have disappeared in a puff of statistics," he added:

> However, it is only on the accurate lunar lines that this disappointing conclusion has been reached. More exciting is the positive conclusion that the megalith builders *were* interested in the motions of the sun and moon, and that they *did* mark the extreme southern solar and lunar positions.[3]

On this basis, Norris encouraged further investigations, cautioning, however, that such tests will involve "years of painstaking data collection."

---

[2]Alvar Ellegård, "[Review of] *Megalithic Astronomy* . . . [by] C. L. N. Ruggles," *Archaeoastronomy: Supplement to the Journal for the History of Astronomy, 16* (1985), S154.

[3]Ray Norris, "Megalithic Observatories in Britain: Real or Imagined?" in C. L. N. Ruggles (ed.), *Records in Stone: Papers in Memory of Alexander Thom* (Cambridge, England, 1988), p. 275.

# Select Bibliography

*Surveys*


Berry, Arthur. *A Short History of Astronomy: From Earliest Times to the Nineteenth Century.* New York: Dover, 1961 republication of the 1898 original.

Cronin, Vincent. *The View from Planet Earth: Man Looks at the Cosmos.* New York: William Morrow, 1981. Includes much information on interactions between astronomy and other areas of culture.

Dreyer, J. L. E. *A History of Astronomy from Thales to Kepler.* New York: Dover, 1953 republication of the 1906 original, which appeared as *A History of the Planetary Systems from Thales to Kepler.* Although dated, remains a valuable presentation.

Duhem, Pierre. *To Save the Phenomena: An Essay on the Idea of a Physical Theory from Plato to Galileo.* Translated by Edmund Doland and Chaninah Maschler. Chicago: Univ. of Chicago Press, 1969. A classic study of this subject. For a critique, see: G. E. R. Lloyd. "Saving the Appearances." *Classical Quarterly,* 28 (1978): 202–22.

Durham, Frank, and Robert D. Purrington. *Frame of the Universe: A History of Physical Cosmology.* New York: Columbia Univ. Press, 1983.

Gingerich, Owen. "Ptolemy, Copernicus, Kepler," *Great Ideas Today 1983* (Chicago: Encyclopaedia Britannica, Inc., 1983): 137–80.

O'Neil, W. M. *Early Astronomy: From Babylonia to Copernicus.* Sydney: Sydney Univ. Press, 1986. A useful recent survey.

Pannekoek, A. *A History of Astronomy.* New York: Dover, 1989 reprint of the 1961 translation of the 1951 Dutch original. Excellent survey.

Toulmin, Stephen, and June Goodfield. *The Fabric of the Heavens: The Development of Astronomy and Dynamics.* New York: Harper & Row, 1961.

*Archaeoastronomy*

Valuable introductory presentations include:

Brown, Peter Lancaster. *Megaliths, Myths and Men: An Introduction to Astro-Archaeology*. New York: Taplinger, 1976.

Hadingham, Evan. *Early Man and the Cosmos*. Norman: Univ. of Oklahoma Press, 1984. Treats both Old and New World materials; includes a useful bibliography.

Hawkins, Gerald. *Beyond Stonehenge*. New York: Harper & Row, 1973.

Hawkins, Gerald. *Stonehenge Decoded*. Garden City, N. Y.: Delta, 1965.

Heggie, D. C. *Megalithic Science: Ancient Mathematics and Astronomy in North-West Europe*. London: Thames and Hudson, 1981. A critical but not unfriendly analysis by a mathematician. Includes megalithic mathematics.

Hoyle, Fred. *On Stonehenge*. San Francisco: W. H. Freeman. 1977.

Krupp, E. C. *Echoes of the Ancient Skies: The Astronomy of Lost Civilizations*. New York: Harper & Row, 1983. Contains a 30 page bibliography.

Krupp, E. C., ed. *In Search of Ancient Astronomies*. Garden City, N. Y.: Doubleday, 1978. Contains chapters by Krupp, the Thoms, Eddy, and Aveni.

Michell, John. *Megalithomania*. Ithaca: Cornell Univ. Press, 1982. Treats the history of the astronomical interpretation of the megalithic sites.

For more advanced treatments, see:

Ruggles, C. L. N., ed. *Records in Stone: Papers in Memory of Alexander Thom*. Cambridge: Cambridge Univ. Press, 1988. Contains 21 papers by experts.

Thom, Alexander. *Megalithic Lunar Observatories*. Oxford: Clarendon, 1971.

Thom, Alexander. *Megalithic Sites in Britain*. Oxford: Clarendon, 1967.

Thom, Alexander, and A. S. Thom. *Megalithic Remains in Britain and Britanny*. Oxford: Clarendon, 1978.

For a survey of pre-Columbian astronomy in Central America, see:

Aveni, Anthony F. *Skywatchers of Ancient Mexico*. Austin: Univ. of Texas Press, 1980. Contains a helpful presentation of the chief celestial appearances.

Two journals are devoted to archaeoastronomy:

*Archaeoastronomy: The Journal of the Center for Archaeoastronomy*.

*Archaeoastronomy: Supplement to the Journal for the History of Astronomy*. In 1979, *Journal for the History of Astronomy*, where a number of important archaeoastronomy papers had appeared, began to devote one issue each year to this subject.

*Ancient, Chiefly Greek, Astronomy*

A highly respected introduction to ancient astronomy, including Egyptian and Babylonian astronomy, is:

Neugebauer, Otto. *The Exact Sciences in Antiquity*. 2d ed. New York: Dover, 1969.

For a far more detailed study, see:

Neugebauer, Otto. *A History of Ancient Mathematical Astronomy*. 3 vols. New York: Springer-Verlag, 1975. See also Neugebauer's *Astronomy and History: Selected Essays*. New York: Springer-Verlag, 1983.

Among books containing reliable information on Greek astronomy are:

Dicks, D. R. *Early Greek Astronomy to Aristotle*. Ithaca: Cornell Univ. Press, 1970.

Evans, James Carl. *The History and Practice of Ancient Astronomy*. Doctoral dissertation, University of Washington, 1983. Contains a discussion of Greek astronomy as well materials Evans developed for an undergraduate course on Greek astronomy.

Heath, Thomas Little. *Aristarchus of Samos, The Ancient Copernicus: A History of Greek Astronomy to Aristarchus together with Aristarchus's Treatise on the Sizes and Distances of the Sun and Moon*. Oxford: Clarendon, 1913. Reprinted by Dover; New York, 1981. Presents Aristarchus's treatise as well as much valuable historical information.

Heath, Thomas Little. *Greek Astronomy*. London: J. M. Dent & Sons, 1932. Consists chiefly of texts; much less detailed than his *Aristarchus,* but extends to Ptolemy.

Hetherington, Norriss S. *Ancient Astronomy and Civilization.* Tucson: Pachart, 1987.

Hodson, F. R., ed. *The Place of Astronomy in the Ancient World.* London: Oxford Univ. Press, 1974.

Pedersen, O., and M. Pihl. *Early Physics and Astronomy.* New York: Science History, 1974.

On Ptolemy:

The best translation of his *Almagest* is:

*Ptolemy's Almagest.* Translated by G. J. Toomer. New York: Springer-Verlag, 1984. Another good translation is that by R. C. Taliaferro in vol. XVI of *Great Books of the Western World.*

Also see:

Evans, James. "On the Function and Probable Origin of Ptolemy's Equant." *American Journal of Physics,* 52 (Dec., 1984): 1080–90. An especially clear and interesting presentation on this important topic.

Newton, Robert R. *The Crime of Claudius Ptolemy.* Baltimore: Johns Hopkins Univ. Press, 1977. In this and other publications, Newton has claimed that Ptolemy derived most of his key ideas from Hipparchus without giving him proper credit. An extended controversy has resulted. See, for example, Noel Swerdlow. "Ptolemy on Trial." *American Scholar,* 48 (1979): 523–31.

Pedersen, Olaf. *A Survey of the Almagest.* Odense: Odense Univ. Press, 1974.

Toomer, G. J. "Ptolemy." In *Dictionary of Scientific Biography,* vol. XI, pp. 186–206. New York: Charles Scribner's Sons, 1975.

For clear, elementary mathematical presentations of Ptolemy's planetary system, see:

Benjamin, F. S., and G. J. Toomer. *Campanus of Novara and Medieval Planetary Theory.* Madison: Univ. of Wisconsin Press, 1971, esp. pp. 39–57.

Neugebauer, Otto. *The Exact Sciences in Antiquity* (listed above), pp. 191–206.

*The Copernican Revolution*

Works of a general nature include:

Christianson, Gale E. *This Wild Abyss: The Story of the Men Who Made Modern Astronomy.* New York: Free Press, 1978. A high level popularization covering the period from Copernicus to Newton.

Koestler, Arthur. *The Sleepwalkers: A History of Man's Changing Vision of the Universe.* New York: Macmillan, 1959. A controversial and entertaining study by a well known author.

Koyré, Alexandre. *From the Closed World to the Infinite Universe.* Baltimore: Johns Hopkins Univ. Press, 1957. Discusses the Copernican "infinitization" of the universe.

Koyré, Alexandre. *The Astronomical Revolution: Copernicus-Kepler-Borelli.* Translated by R. E. W. Maddison. Paris: Hermann, 1973. Koyré was one of the most influential historians of science of this century.

Kuhn, Thomas S. *The Copernican Revolution.* Cambridge: Harvard Univ. Press, 1957.

The four main figures in astronomy from 1500 to 1640 were Copernicus, Brahe, Kepler, and Galileo. On them, see the articles in the *Dictionary of Scientific Biography* and the publications listed below.

Copernicus: The most authoritative source is: Noel M. Swerdlow and O. Neugebauer. *Mathematical Astronomy in Copernicus's De Revolutionibus.* 2 vols. New York: Springer-Verlag, 1984. Copernicus's *De revolutionibus* is available in two highly regarded English translations, those by Edward Rosen and by A. M. Duncan. For a valuable collection of articles, see: Robert Westman, ed. *The Copernican Achievement.* Berkeley: Univ. of California Press, 1975. Many useful primary documents appear in: Edward Rosen. *Copernicus and the Scientific Revolution.* Malabar, Florida: Robert E. Krieger, 1984.

Brahe: The standard if somewhat dated biography is: J. L. E. Dreyer. *Tycho Brahe.* New York: Dover, 1963 reprint of the 1890 original.

Kepler: The standard biography is Max Caspar. *Kepler.*

Translated by C. D. Hellman. London: Abelard-Schuman, 1959. Gradually various writings by Kepler are appearing in English translations. See also: Arthur Beer and Peter Beer, eds. *Kepler: Four Hundred Years*. Oxford: Pergamon, 1975.

Galileo: Among many biographies of Galileo, the most highly regarded is: Stillman Drake. *Galileo at Work: His Scientific Biography*. Chicago: Univ. of Chicago Press, 1978. Drake and others have translated many of Galileo's most important writings. For a valuable collection of essays, see: Ernan McMullin, ed. *Galileo: Man of Science*. New York: Basic Books, 1967.

Miscellaneous relevant studies:

Burtt, E. A. *The Metaphysical Foundations of Modern Physical Science*. New York: Doubleday, 1954. A classic study of the interrelations between philosophy and the new science.

Cohen, I. B. *The Birth of a New Physics*. 2d ed. New York: W. W. Norton, 1985. An excellent elementary survey of mechanics during the scientific revolution.

Crowe, Michael J. *The Extraterrestrial Life Debate 1750–1900: The Idea of a Plurality of Worlds from Kant to Lowell*. Cambridge: Cambridge Univ. Press, 1986. See Chapter I.

Dick, Steven J. *The Plurality of Worlds: The Origins of the Extraterrestrial Life Debate from Democritus to Kant*. Cambridge: Cambridge Univ. Press, 1982. A fine study that sheds much light on the astronomical revolution.

King, Henry C. *The History of the Telescope*. New York: Dover, 1979 reprint of the 1955 original. The standard history of the telescope.

Michel, Paul-Henri. *The Cosmology of Giordano Bruno*. Translated by R. E. W. Maddison. Paris: Hermann, 1973. A fine study of this controversial figure.

Nicolson, Marjorie. *Science and Imagination*. Ithaca: Cornell Univ. Press, 1956. A valuable collection of articles by a pioneer in the study of the interactions between science and literature.

Van Helden, Albert. *Measuring the Universe: Cosmic Dimensions from Aristarchus to Halley*. Chicago: Univ. of Chicago Press, 1984. An important study of a crucial topic.

# Name Index


Adam 186, 189
Addison, Joseph 190
Al-Battani 129
Al-Bitruji 128
Anaximander 116
Anaximenes 116
Apollonius of Perga 31
Archimedes of Syracuse 28, 30
Aristarchus of Samos 27–30, 89
Aristotle 26–7, 53, 73, 84, 122–3, 126, 133, 139
Ashworth, William viii
Atkinson, R. J. C. 211, 213
Aubrey, John 202
Averroes 73–4, 76
Bacon, Francis vii
Bacon, Roger 76
Baity, Elizabeth 219
Bellarmine, Cardinal Robert 78, 83–4, 158, 159
Benjamin, F. S. 46
Bernard of Verdun 76
Blake, William 202
Brahe, Tycho 137–46, 152, 153, 156, 195
Brummer, Jamie viii
Bruno, Giordano 175, 177
Callippus of Cyzicus 25–6
Capella, Martianus 130
Casper, Barry M. 37
Cazre, Pierre de 180–1
Chanut, Pierre 181
Charleton, Walter 202
Christ, Jesus 15, 100, 152, 174, 181, 189
Cicero, Marcus Tullius 108, 118
Cleanthes 30
Clement VII, Pope 87
Constable, John 202
Copernicus, Nicholas v, 30, 77, 80–1, 85–136, 140–2, 152, 158, 168, 174, 175, 192–3, 194–7
Coronel, Luiz 76
Crosson, Frederick J. viii
Dantiscus, John 87
Davenant, William 182
De Morgan, Augustus 194
Democritus of Abdera 116
Descartes, René 181–2
Dilthey, Wilhelm vi
Divini, Eustace 171
Dobrzycki, Jerzy 102
Donne, John 177–9
Dostoyevsky, Feodor 149
Drake, Stillman 160
Dreyer, J. L. E. iii, 147–8
Drummond, William 180
Duhem, Pierre 69, 76
Duke, Edward 202
Ecphantus 108, 118
Ellegård, Alvar 220
Empedocles 116
Eratosthenes of Alexandria 30–1
Euclid of Alexandria 45, 128
Eudoxus of Cnidus 23–5, 28, 175
Evans, James viii
Fabricius, David 171
Flaumenhaft, Harvey viii
Fontana, Francesco 171
Foscarini, Paolo Antonio 83
Frederick II (Danish King) 138, 143
Galilei, Galileo 77, 82, 152, 157–72, 177, 195
Galilei, Vincenzio 157
Gassendi, Pierre 171, 180
Geminus 70–1
Geoffrey of Monmouth 202
Giese, Tiedemann 88, 106
Goddu, André iv, viii, 69
Goethe, Johann Wolfgang von 86, 194

Gottesman, Stephen viii
Grafton, John W. viii
Hatch, Robert A. viii
Hawkins, Gerald 209–11, 213, 214, 218
Heraclides of Pontus 27, 71, 108, 118
Heraclitus of Ephesus 27, 116
Herbert, George 179–80
Hermes Trismegistus 133
Hevelius, Johannes 171
Hicetas of Syracuse 108, 118
Hipparchus of Nicaea 31, 32, 41–2, 46, 105, 175
Hoyle, Fred 210–12
Huygens, Christiaan 170, 189
Jones, Inigo 202
Joshua 174
Kant, Immanuel 192–3
Kepler, Johannes vi, 77, 82, 88, 143, 147–56, 158, 160, 168, 177, 195
Kircher, Athanasius 172
Koestler, Arthur 163–4
Kopal, Zdenek 164
Lactantius 109
Laplace, Pierre Simon 193
Leibniz, Gottfried Wilhelm 189
Leo X, Pope 109
Leucippus 116
Lipperhey, Hans 158
Lockyer, J. Norman 203–4
Luther, Martin vi, 79–80, 174
Maestlin, Michael 147, 149, 152
Maimonides, Moses 74–5
Medici, Grand Duke Cosimo II de 158, 159
Melanchthon, Philip 80–4, 174
Milton, John 195–8
Montaigne, Michel de 175
More, Henry 181
Neugebauer, Otto 195–6
Newham, C. A. 213–15, 218
Nicholas of Cusa 76
Noer, Richard J. 37
Norris, Ray 220
Novara, Domenico Maria di 87
Osiander, Andreas 77–9, 81, 84, 88, 103–4
Pascal, Blaise 182–5
Paul III, Pope 88, 105–8
Paul of Middelburg 109
Paul V, Pope 159
Penrose, F. C. 203
Peter of Padua 76
Petreius, Johannes 102
Petrie, Flinders 202
Peucer, Caspar 81
Philolaus of Croton 108, 119
Plato 23, 26, 102, 119, 128
Plutarch of Chaeronea 30, 112
Polanyi, Michael 196–7
Pope, Alexander 190–1
Proclus 72–3
Ptolemy, Claudius 23, 31, 32, 34, 40, 45–68, 72, 74, 75, 89–91, 94, 95, 98, 112, 115, 123–4, 129–30, 135, 136, 142, 144–6, 152, 175, 194
Pye, Samuel 192
Raphael 182–4
Reinhold, Erasmus 81
Rheticus, Georg Joachim 87–8
Riccioli, Giovanni Battista 171
Rosen, Edward 102
Rudolph II, Emperor 143
Ruggles, Clive 220
Rushd, Ibn 129
Scheiner, Christopher 169, 171
Schönberg, Cardinal Nicholas 87, 104–6
Shakespeare, William 176
Simplicius 23
Sloan, Phillip R. viii
Smith, John 202
Sophocles 133
Stukeley, William 202
Syrus 53
Theodoric of Regen 105
Thom, Alexander 213, 215–20
Thom, Alexander Strang 219
Thom, Archibald 216, 219

Thomas Aquinas 75–6
Toomer, G. J. 46, 53
Traherne, Thomas 185
Turner, Joseph 202
Urban VIII, Pope 159
Vergil 124
Watzenrode, Lucas 86–7
Widmanstadt, Johann 87
Wilson, Patrick A. viii
Xenophanes of Colophon 116
Young, Edward 191–2

A CATALOG OF SELECTED

# DOVER BOOKS

IN ALL FIELDS OF INTEREST

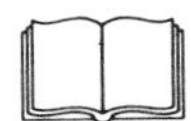

THE BOOK OF BEASTS: Being a Translation from a Latin Bestiary of the Twelfth Century, T. H. White. Wonderful catalog of real and fanciful beasts: manticore, griffin, phoenix, amphivius, jaculus, many more. White's witty erudite commentary on scientific, historical aspects enhances fascinating glimpse of medieval mind. Illustrated. 296pp. 5⅜ × 8¼. (Available in U.S. only) 24609-4 Pa. $7.95

FRANK LLOYD WRIGHT: Architecture and Nature with 160 Illustrations, Donald Hoffmann. Profusely illustrated study of influence of nature—especially prairie—on Wright's designs for Fallingwater, Robie House, Guggenheim Museum, other masterpieces. 96pp. 9¼ × 10¾. 25098-9 Pa. $8.95

LIMBERT ARTS AND CRAFTS FURNITURE: The Complete 1903 Catalog, Charles P. Limbert and Company. Rare catalog depicting 188 pieces of Mission-style furniture: fold-down tables and desks, bookcases, library and octagonal tables, chairs, more. Descriptive captions. 80pp. 9⅜ × 12¼. 27120-X Pa. $6.95

YEARS WITH FRANK LLOYD WRIGHT: Apprentice to Genius, Edgar Tafel. Insightful memoir by a former apprentice presents a revealing portrait of Wright the man, the inspired teacher, the greatest American architect. 372 black-and-white illustrations. Preface. Index. vi + 228pp. 8¼ × 11. 24801-1 Pa. $10.95

THE STORY OF KING ARTHUR AND HIS KNIGHTS, Howard Pyle. Enchanting version of King Arthur fable has delighted generations with imaginative narratives of exciting adventures and unforgettable illustrations by the author. 41 illustrations. xviii + 313pp. 6⅛ × 9¼. 21445-1 Pa. $6.95

THE GODS OF THE EGYPTIANS, E. A. Wallis Budge. Thorough coverage of numerous gods of ancient Egypt by foremost Egyptologist. Information on evolution of cults, rites and gods; the cult of Osiris; the Book of the Dead and its rites; the sacred animals and birds; Heaven and Hell; and more. 956pp. 6⅛ × 9¼. 22055-9, 22056-7 Pa., Two-vol. set $22.90

A THEOLOGICO-POLITICAL TREATISE, Benedict Spinoza. Also contains unfinished *Political Treatise*. Great classic on religious liberty, theory of government on common consent. R. Elwes translation. Total of 421pp. 5⅜ × 8½. 20249-6 Pa. $7.95

INCIDENTS OF TRAVEL IN CENTRAL AMERICA, CHIAPAS, AND YUCATAN, John L. Stephens. Almost single-handed discovery of Maya culture; exploration of ruined cities, monuments, temples; customs of Indians. 115 drawings. 892pp. 5⅜ × 8½. 22404-X, 22405-8 Pa., Two-vol. set $17.90

LOS CAPRICHOS, Francisco Goya. 80 plates of wild, grotesque monsters and caricatures. Prado manuscript included. 183pp. 6⅝ × 9⅝. 22384-1 Pa. $6.95

AUTOBIOGRAPHY: The Story of My Experiments with Truth, Mohandas K. Gandhi. Not hagiography, but Gandhi in his own words. Boyhood, legal studies, purification, the growth of the Satyagraha (nonviolent protest) movement. Critical, inspiring work of the man who freed India. 480pp. 5⅜ × 8½. (Available in U.S. only) 24593-4 Pa. $6.95

ILLUSTRATED DICTIONARY OF HISTORIC ARCHITECTURE, edited by Cyril M. Harris. Extraordinary compendium of clear, concise definitions for over 5,000 important architectural terms complemented by over 2,000 line drawings. Covers full spectrum of architecture from ancient ruins to 20th-century Modernism. Preface. 592pp. 7½ × 9⅝. 24444-X Pa. $15.95

THE NIGHT BEFORE CHRISTMAS, Clement C. Moore. Full text, and woodcuts from original 1848 book. Also critical, historical material. 19 illustrations. 40pp. 4⅝ × 6. 22797-9 Pa. $2.50

THE LESSON OF JAPANESE ARCHITECTURE: 165 Photographs, Jiro Harada. Memorable gallery of 165 photographs taken in the 1930s of exquisite Japanese homes of the well-to-do and historic buildings. 13 line diagrams. 192pp. 8⅜ × 11¼. 24778-3 Pa. $10.95

THE AUTOBIOGRAPHY OF CHARLES DARWIN AND SELECTED LETTERS, edited by Francis Darwin. The fascinating life of eccentric genius composed of an intimate memoir by Darwin (intended for his children); commentary by his son, Francis; hundreds of fragments from notebooks, journals, papers; and letters to and from Lyell, Hooker, Huxley, Wallace and Henslow. xi + 365pp. 5⅜ × 8. 20479-0 Pa. $6.95

WONDERS OF THE SKY: Observing Rainbows, Comets, Eclipses, the Stars and Other Phenomena, Fred Schaaf. Charming, easy-to-read poetic guide to all manner of celestial events visible to the naked eye. Mock suns, glories, Belt of Venus, more. Illustrated. 299pp. 5¼ × 8¼. 24402-4 Pa. $8.95

BURNHAM'S CELESTIAL HANDBOOK, Robert Burnham, Jr. Thorough guide to the stars beyond our solar system. Exhaustive treatment. Alphabetical by constellation: Andromeda to Cetus in Vol. 1; Chamaeleon to Orion in Vol. 2; and Pavo to Vulpecula in Vol. 3. Hundreds of illustrations. Index in Vol. 3. 2,000pp. 6⅛ × 9¼. 23567-X, 23568-8, 23673-0 Pa., Three-vol. set $41.85

STAR NAMES: Their Lore and Meaning, Richard Hinckley Allen. Fascinating history of names various cultures have given to constellations and literary and folkloristic uses that have been made of stars. Indexes to subjects. Arabic and Greek names. Biblical references. Bibliography. 563pp. 5⅜ × 8½. 21079-0 Pa. $9.95

THIRTY YEARS THAT SHOOK PHYSICS: The Story of Quantum Theory, George Gamow. Lucid, accessible introduction to influential theory of energy and matter. Careful explanations of Dirac's anti-particles, Bohr's model of the atom, much more. 12 plates. Numerous drawings. 240pp. 5⅜ × 8½. 24895-X Pa. $6.95

CHINESE DOMESTIC FURNITURE IN PHOTOGRAPHS AND MEASURED DRAWINGS, Gustav Ecke. A rare volume, now affordably priced for antique collectors, furniture buffs and art historians. Detailed review of styles ranging from early Shang to late Ming. Unabridged republication. 161 black-and-white drawings, photos. Total of 224pp. 8⅜ × 11¼. (Available in U.S. only) 25171-3 Pa. $14.95

VINCENT VAN GOGH: A Biography, Julius Meier-Graefe. Dynamic, penetrating study of artist's life, relationship with brother, Theo, painting techniques, travels, more. Readable, engrossing. 160pp. 5⅜ × 8½. (Available in U.S. only) 25253-1 Pa. $4.95

HOW TO WRITE, Gertrude Stein. Gertrude Stein claimed anyone could understand her unconventional writing—here are clues to help. Fascinating improvisations, language experiments, explanations illuminate Stein's craft and the art of writing. Total of 414pp. 4⅝ × 6⅜. 23144-5 Pa. $6.95

ADVENTURES AT SEA IN THE GREAT AGE OF SAIL: Five Firsthand Narratives, edited by Elliot Snow. Rare true accounts of exploration, whaling, shipwreck, fierce natives, trade, shipboard life, more. 33 illustrations. Introduction. 353pp. 5⅜ × 8½. 25177-2 Pa. $9.95

THE HERBAL OR GENERAL HISTORY OF PLANTS, John Gerard. Classic descriptions of about 2,850 plants—with over 2,700 illustrations—includes Latin and English names, physical descriptions, varieties, time and place of growth, more. 2,706 illustrations. xlv + 1,678pp. 8½ × 12¼. 23147-X Cloth. $89.95

DOROTHY AND THE WIZARD IN OZ, L. Frank Baum. Dorothy and the Wizard visit the center of the Earth, where people are vegetables, glass houses grow and Oz characters reappear. Classic sequel to *Wizard of Oz*. 256pp. 5⅜ × 8. 24714-7 Pa. $5.95

SONGS OF EXPERIENCE: Facsimile Reproduction with 26 Plates in Full Color, William Blake. This facsimile of Blake's original "Illuminated Book" reproduces 26 full-color plates from a rare 1826 edition. Includes "The Tyger," "London," "Holy Thursday," and other immortal poems. 26 color plates. Printed text of poems. 48pp. 5¼ × 7. 24636-1 Pa. $3.95

SONGS OF INNOCENCE, William Blake. The first and most popular of Blake's famous "Illuminated Books," in a facsimile edition reproducing all 31 brightly colored plates. Additional printed text of each poem. 64pp. 5¼ × 7. 22764-2 Pa. $3.95

PRECIOUS STONES, Max Bauer. Classic, thorough study of diamonds, rubies, emeralds, garnets, etc.: physical character, occurrence, properties, use, similar topics. 20 plates, 8 in color. 94 figures. 659pp. 6⅛ × 9¼. 21910-0, 21911-9 Pa., Two-vol. set $21.90

ENCYCLOPEDIA OF VICTORIAN NEEDLEWORK, S. F. A. Caulfeild and Blanche Saward. Full, precise descriptions of stitches, techniques for dozens of needlecrafts—most exhaustive reference of its kind. Over 800 figures. Total of 679pp. 8⅜ × 11. 22800-2, 22801-0 Pa., Two-vol. set $26.90

THE MARVELOUS LAND OF OZ, L. Frank Baum. Second Oz book, the Scarecrow and Tin Woodman are back with hero named Tip, Oz magic. 136 illustrations. 287pp. 5⅜ × 8½. 20692-0 Pa. $5.95

WILD FOWL DECOYS, Joel Barber. Basic book on the subject, by foremost authority and collector. Reveals history of decoy making and rigging, place in American culture, different kinds of decoys, how to make them, and how to use them. 140 plates. 156pp. 7⅞ × 10¾. 20011-6 Pa. $14.95

HISTORY OF LACE, Mrs. Bury Palliser. Definitive, profusely illustrated chronicle of lace from earliest times to late 19th century. Laces of Italy, Greece, England, France, Belgium, etc. Landmark of needlework scholarship. 266 illustrations. 672pp. 6⅛ × 9¼. 24742-2 Pa. $16.95

ILLUSTRATED GUIDE TO SHAKER FURNITURE, Robert Meader. All furniture and appurtenances, with much on unknown local styles. 235 photos. 146pp. 9 × 12. 22819-3 Pa. $9.95

WHALE SHIPS AND WHALING: A Pictorial Survey, George Francis Dow. Over 200 vintage engravings, drawings, photographs of barks, brigs, cutters, other vessels. Also harpoons, lances, whaling guns, many other artifacts. Comprehensive text by foremost authority. 207 black-and-white illustrations. 288pp. 6 × 9. 24808-9 Pa. $9.95

THE BERTRAMS, Anthony Trollope. Powerful portrayal of blind self-will and thwarted ambition includes one of Trollope's most heartrending love stories. 497pp. 5⅜ × 8½. 25119-5 Pa. $9.95

ADVENTURES WITH A HAND LENS, Richard Headstrom. Clearly written guide to observing and studying flowers and grasses, fish scales, moth and insect wings, egg cases, buds, feathers, seeds, leaf scars, moss, molds, ferns, common crystals, etc.—all with an ordinary, inexpensive magnifying glass. 209 exact line drawings aid in your discoveries. 220pp. 5⅜ × 8½. 23330-8 Pa. $5.95

RODIN ON ART AND ARTISTS, Auguste Rodin. Great sculptor's candid, wide-ranging comments on meaning of art; great artists; relation of sculpture to poetry, painting, music; philosophy of life, more. 76 superb black-and-white illustrations of Rodin's sculpture, drawings and prints. 119pp. 8⅜ × 11¼. 24487-3 Pa. $7.95

FIFTY CLASSIC FRENCH FILMS, 1912–1982: A Pictorial Record, Anthony Slide. Memorable stills from Grand Illusion, Beauty and the Beast, Hiroshima, Mon Amour, many more. Credits, plot synopses, reviews, etc. 160pp. 8¼ × 11. 25256-6 Pa. $11.95

THE PRINCIPLES OF PSYCHOLOGY, William James. Famous long course complete, unabridged. Stream of thought, time perception, memory, experimental methods; great work decades ahead of its time. 94 figures. 1,391pp. 5⅜ × 8½. 20381-6, 20382-4 Pa., Two-vol. set $25.90

BODIES IN A BOOKSHOP, R. T. Campbell. Challenging mystery of blackmail and murder with ingenious plot and superbly drawn characters. In the best tradition of British suspense fiction. 192pp. 5⅜ × 8½. 24720-1 Pa. $5.95

CALLAS: Portrait of a Prima Donna, George Jellinek. Renowned commentator on the musical scene chronicles incredible career and life of the most controversial, fascinating, influential operatic personality of our time. 64 black-and-white photographs. 416pp. 5⅜ × 8¼. 25047-4 Pa. $8.95

GEOMETRY, RELATIVITY AND THE FOURTH DIMENSION, Rudolph Rucker. Exposition of fourth dimension, concepts of relativity as Flatland characters continue adventures. Popular, easily followed yet accurate, profound. 141 illustrations. 133pp. 5⅜ × 8½. 23400-2 Pa. $4.95

HOUSEHOLD STORIES BY THE BROTHERS GRIMM, with pictures by Walter Crane. 53 classic stories—Rumpelstiltskin, Rapunzel, Hansel and Gretel, the Fisherman and his Wife, Snow White, Tom Thumb, Sleeping Beauty, Cinderella, and so much more—lavishly illustrated with original 19th-century drawings. 114 illustrations. x + 269pp. 5⅜ × 8½. 21080-4 Pa. $4.95

SUNDIALS, Albert Waugh. Far and away the best, most thorough coverage of ideas, mathematics concerned, types, construction, adjusting anywhere. Over 100 illustrations. 230pp. 5⅜ × 8½. 22947-5 Pa. $5.95

PICTURE HISTORY OF THE NORMANDIE: With 190 Illustrations, Frank O. Braynard. Full story of legendary French ocean liner: Art Deco interiors, design innovations, furnishings, celebrities, maiden voyage, tragic fire, much more. Extensive text. 144pp. 8⅞ × 11¾. 25257-4 Pa. $11.95

THE FIRST AMERICAN COOKBOOK: A Facsimile of "American Cookery," 1796, Amelia Simmons. Facsimile of the first American-written cookbook published in the United States contains authentic recipes for colonial favorites—pumpkin pudding, winter squash pudding, spruce beer, Indian slapjacks, and more. Introductory Essay and Glossary of colonial cooking terms. 80pp. 5⅜ × 8½. 24710-4 Pa. $3.50

101 PUZZLES IN THOUGHT AND LOGIC, C. R. Wylie, Jr. Solve murders and robberies, find out which fishermen are liars, how a blind man could possibly identify a color—purely by your own reasoning! 107pp. 5⅜ × 8½. 20367-0 Pa. $2.95

ANCIENT EGYPTIAN MYTHS AND LEGENDS, Lewis Spence. Examines animism, totemism, fetishism, creation myths, deities, alchemy, art and magic, other topics. Over 50 illustrations. 432pp. 5⅜ × 8½. 26525-0 Pa. $8.95

ANTHROPOLOGY AND MODERN LIFE, Franz Boas. Great anthropologist's classic treatise on race and culture. Introduction by Ruth Bunzel. Only inexpensive paperback edition. 255pp. 5⅜ × 8½. 25245-0 Pa. $7.95

THE TALE OF PETER RABBIT, Beatrix Potter. The inimitable Peter's terrifying adventure in Mr. McGregor's garden, with all 27 wonderful, full-color Potter illustrations. 55pp. 4¼ × 5½. 22827-4 Pa. $1.75

THREE PROPHETIC SCIENCE FICTION NOVELS, H. G. Wells. *When the Sleeper Wakes, A Story of the Days to Come* and *The Time Machine* (full version). 335pp. 5⅜ × 8½. (Available in U.S. only) 20605-X Pa. $8.95

APICIUS COOKERY AND DINING IN IMPERIAL ROME, edited and translated by Joseph Dommers Vehling. Oldest known cookbook in existence offers readers a clear picture of what foods Romans ate, how they prepared them, etc. 49 illustrations. 301pp. 6⅛ × 9¼. 23563-7 Pa. $8.95

SHAKESPEARE LEXICON AND QUOTATION DICTIONARY, Alexander Schmidt. Full definitions, locations, shades of meaning of every word in plays and poems. More than 50,000 exact quotations. 1,485pp. 6½ × 9¼.
22726-X, 22727-8 Pa., Two-vol. set $31.90

THE WORLD'S GREAT SPEECHES, edited by Lewis Copeland and Lawrence W. Lamm. Vast collection of 278 speeches from Greeks to 1970. Powerful and effective models; unique look at history. 842pp. 5⅜ × 8½. 20468-5 Pa. $12.95

THE BLUE FAIRY BOOK, Andrew Lang. The first, most famous collection, with many familiar tales: Little Red Riding Hood, Aladdin and the Wonderful Lamp, Puss in Boots, Sleeping Beauty, Hansel and Gretel, Rumpelstiltskin; 37 in all. 138 illustrations. 390pp. 5⅜ × 8½. 21437-0 Pa. $6.95

THE STORY OF THE CHAMPIONS OF THE ROUND TABLE, Howard Pyle. Sir Launcelot, Sir Tristram and Sir Percival in spirited adventures of love and triumph retold in Pyle's inimitable style. 50 drawings, 31 full-page. xviii + 329pp. 6½ × 9¼. 21883-X Pa. $7.95

THE MYTHS OF THE NORTH AMERICAN INDIANS, Lewis Spence. Myths and legends of the Algonquins, Iroquois, Pawnees and Sioux with comprehensive historical and ethnological commentary. 36 illustrations. 5⅜ × 8½. 25967-6 Pa. $8.95

GREAT DINOSAUR HUNTERS AND THEIR DISCOVERIES, Edwin H. Colbert. Fascinating, lavishly illustrated chronicle of dinosaur research, 1820s to 1960. Achievements of Cope, Marsh, Brown, Buckland, Mantell, Huxley, many others. 384pp. 5¼ × 8¼. 24701-5 Pa. $8.95

THE TASTEMAKERS, Russell Lynes. Informal, illustrated social history of American taste 1850s–1950s. First popularized categories Highbrow, Lowbrow, Middlebrow. 129 illustrations. New (1979) afterword. 384pp. 6 × 9. 23993-4 Pa. $8.95

NORTH AMERICAN INDIAN LIFE: Customs and Traditions of 23 Tribes, Elsie Clews Parsons (ed.). 27 fictionalized essays by noted anthropologists examine religion, customs, government, additional facets of life among the Winnebago, Crow, Zuni, Eskimo, other tribes. 480pp. 6⅛ × 9¼. 27377-6 Pa. $10.95

AUTHENTIC VICTORIAN DECORATION AND ORNAMENTATION IN FULL COLOR: 46 Plates from "Studies in Design," Christopher Dresser. Superb full-color lithographs reproduced from rare original portfolio of a major Victorian designer. 48pp. 9¼ × 12¼. 25083-0 Pa. $7.95

PRIMITIVE ART, Franz Boas. Remains the best text ever prepared on subject, thoroughly discussing Indian, African, Asian, Australian, and, especially, Northern American primitive art. Over 950 illustrations show ceramics, masks, totem poles, weapons, textiles, paintings, much more. 376pp. 5⅜ × 8. 20025-6 Pa. $8.95

SIDELIGHTS ON RELATIVITY, Albert Einstein. Unabridged republication of two lectures delivered by the great physicist in 1920–21. *Ether and Relativity* and *Geometry and Experience*. Elegant ideas in nonmathematical form, accessible to intelligent layman. vi + 56pp. 5⅜ × 8½. 24511-X Pa. $3.95

THE WIT AND HUMOR OF OSCAR WILDE, edited by Alvin Redman. More than 1,000 ripostes, paradoxes, wisecracks: Work is the curse of the drinking classes, I can resist everything except temptation, etc. 258pp. 5⅜ × 8½. 20602-5 Pa. $4.95

ADVENTURES WITH A MICROSCOPE, Richard Headstrom. 59 adventures with clothing fibers, protozoa, ferns and lichens, roots and leaves, much more. 142 illustrations. 232pp. 5⅜ × 8½. 23471-1 Pa. $4.95

PLANTS OF THE BIBLE, Harold N. Moldenke and Alma L. Moldenke. Standard reference to all 230 plants mentioned in Scriptures. Latin name, biblical reference, uses, modern identity, much more. Unsurpassed encyclopedic resource for scholars, botanists, nature lovers, students of Bible. Bibliography. Indexes. 123 black-and-white illustrations. 384pp. 6 × 9. 25069-5 Pa. $9.95

FAMOUS AMERICAN WOMEN: A Biographical Dictionary from Colonial Times to the Present, Robert McHenry, ed. From Pocahontas to Rosa Parks, 1,035 distinguished American women documented in separate biographical entries. Accurate, up-to-date data, numerous categories, spans 400 years. Indices. 493pp. 6½ × 9¼. 24523-3 Pa. $11.95

THE FABULOUS INTERIORS OF THE GREAT OCEAN LINERS IN HISTORIC PHOTOGRAPHS, William H. Miller, Jr. Some 200 superb photographs capture exquisite interiors of world's great "floating palaces"—1890s to 1980s: *Titanic, Ile de France, Queen Elizabeth, United States, Europa,* more. Approx. 200 black-and-white photographs. Captions. Text. Introduction. 160pp. 8⅞ × 11¾. 24756-2 Pa. $10.95

THE GREAT LUXURY LINERS, 1927–1954: A Photographic Record, William H. Miller, Jr. Nostalgic tribute to heyday of ocean liners. 186 photos of *Ile de France, Normandie, Leviathan, Queen Elizabeth, United States,* many others. Interior and exterior views. Introduction. Captions. 160pp. 9 × 12. 24056-8 Pa. $12.95

A NATURAL HISTORY OF THE DUCKS, John Charles Phillips. Great landmark of ornithology offers complete detailed coverage of nearly 200 species and subspecies of ducks: gadwall, sheldrake, merganser, pintail, many more. 74 full-color plates, 102 black-and-white. Bibliography. Total of 1,920pp. 8⅝ × 11¼. 25141-1, 25142-X Cloth., Two-vol. set $100.00

THE COMPLETE "MASTERS OF THE POSTER": All 256 Color Plates from "Les Maîtres de l'Affiche", Stanley Appelbaum (ed.). The most famous compilation ever made of the art of the great age of the poster, featuring works by Chéret, Steinlen, Toulouse-Lautrec, nearly 100 other artists. One poster per page. 272pp. 9¼ × 12¼. 26309-6 Pa. $29.95

THE TEN BOOKS OF ARCHITECTURE: The 1755 Leoni Edition, Leon Battista Alberti. Rare classic helped introduce the glories of ancient architecture to the Renaissance. 68 black-and-white plates. 336pp. 8⅜ × 11¼. 25239-6 Pa. $14.95

MISS MACKENZIE, Anthony Trollope. Minor masterpieces by Victorian master unmasks many truths about life in 19th-century England. First inexpensive edition in years. 392pp. 5⅜ × 8½. 25201-9 Pa. $8.95

THE RIME OF THE ANCIENT MARINER, Gustave Doré, Samuel Taylor Coleridge. Dramatic engravings considered by many to be his greatest work. The terrifying space of the open sea, the storms and whirlpools of an unknown ocean, the ice of Antarctica, more—all rendered in a powerful, chilling manner. Full text. 38 plates. 77pp. 9¼ × 12. 22305-1 Pa. $4.95

THE EXPEDITIONS OF ZEBULON MONTGOMERY PIKE, Zebulon Montgomery Pike. Fascinating firsthand accounts (1805–6) of exploration of Mississippi River, Indian wars, capture by Spanish dragoons, much more. 1,088pp. 5⅜ × 8½. 25254-X, 25255-8 Pa., Two-vol. set $25.90